PREFACE

I am pleased to offer the Seventh Edition of MULTIPLE-CHOICE & FREE-RESPONSE QUESTIONS IN PREPARATION FOR THE AP CALCULUS (BC) EXAMINATION.

The book has been prepared with one purpose in mind: to enhance your preparation for the Advanced Placement Calculus BC Examination. It should be made clear that you will have to respond to both multiple-choice and free-response questions on the examination.

This 7th edition consists of six sample examinations, each with forty-five multiple-choice questions and six free-response questions. Section I Part A consists of 28 multiple-choice questions to be answered without the use of a calculator. Section I Part B consists of 17 multiple-choice questions, designed with graphing calculators in mind, and contains some questions for which this technology is required. Section II Part A consists of three free-response questions and contains some questions or parts of questions that may require the use of a graphing calculator. Section II Part B consists of three free-response questions for which you may not use a calculator.

In this book, you have an opportunity to take six examinations under conditions that simulate those in an actual test administration. Workspace for each question has been provided for your calculations. You should allow 55 minutes for Section I Part A, 50 minutes for Section I part B, 45 minutes for Section II Part A, and 45 minutes for Section II part B. Reference and review formulas are provided at the end of the book. However, these formulas are not provided on the actual AP examination.

By completing all six examinations, you will be better able to identify your strengths and address your weaknesses. Upon completion of this book, you will then be able to approach the Advanced Placement Calculus BC Examination with increased confidence.

I wish to thank Bob Byrne of St. Thomas Aquinas HS (Fort Lauderdale, FL) for sharing ideas and making valuable suggestions which have been incorporated in this edition. A special thanks to Lin McMullin for contributing many graphing calculator, conceptual, and free-response questions which have been included in this edition. A special debt of gratitude is due to my wife for her encouragement and patience. I also wish to thank Tammi Pruzansky of D&S Marketing Systems, Inc., for preparing the graphs and typesetting the entire manuscript.

Any errors found in this book are solely the responsibility of the author.

All communications concerning this book should be addressed to:

D & S Marketing Systems, Inc.
1205 38th Street
Brooklyn, NY 11218
www.dsmarketing.com

TABLE OF CONTENTS

The AP Calculus Examination

How, not only to Survive, but to Prevail ...

by Lin McMullin

The AP Calculus exam is the cumulation of all of the years you've spent in high school studying mathematics. It all led up to this. The calculus you study in the last year completes the prior years of preparation. If you are reading this at the beginning of the year, keep these things in mind as you go through the year. If you are reading this only a few weeks before the test, think back and see how these things fit together.

Everything in calculus, and mathematics in general, is best understood verbally, numerically, analytically (that is, through the use of equations and symbols) and graphically. Look at everything from these perspectives. Look at the relationships among them – how the same idea shows up in words, in equations, in numbers, and in graphs.

For example: numerically a linear function is one which when written as a table of values, regular changes in the x-values produce regular changes in the y-values. Graphically a linear function has a graph that is a straight line. Analytically it is one whose equation can be written as $y = mx + b$. And the three ways are interrelated. The ratio of the changes in the table is the number m in the equation. The graph can be drawn using the number m by going up and over from one point to the next. The idea of the slope as "rise over run" expresses this verbally. Everything in mathematics and in the calculus works that way.

Learn the concepts — the exam emphasizes concepts

Learn the procedures and formulae — even though the concepts are more important than the computations you still have to do the computations. Like it or not, learn to do the algebra, the arithmetic and the graphs.

Learn to be methodical — work neatly and carefully all year.

Think about what you are doing. Watch yourself work. It is natural to concentrate on the material you know and can do, but you need to concentrate on the things you do not (yet) know how to do. You can learn much from your mistakes. Look at a wrong answer as signal that you need to learn more about that topic.

Reviewing for the Exam

In the few weeks before the AP Exam you will need to review what you have studied, firm up what you have learned, work on your areas of weakness, and yes, memorize some formulas. You also need to prepare for the exam itself by learning what kinds of questions will be asked and how to best answer them. Specifically:

- Understand the format of the exams. (See page iii). Know how your knowledge will be tested.

- STUDY WHAT YOU DO NOT KNOW. That may seem obvious, but many people enjoy getting the right answers so much that they only review the stuff they know. The time to concentrate on what you know is when you are taking the test.

- Practice writing free-response answers. The College Board publishes copies of student answers from past years. If your teacher has some of these, look at them and learn what is expected and what is not needed.

- Plan your review carefully. Don't try to cram the weekend before the exam. The day before the test: relax, get psyched, and get a good night's sleep. On the day of the test eat a good breakfast. The test is grueling, even though you're up for it. Bring a snack for the brief break between the multiple-choice and free-response sections.

Calculators

The reason calculators are so important in learning mathematics is that they allow you do the graphical and numerical work easily, quickly and accurately. You should use your calculator all year, on homework, tests and when studying. Learn how to use it efficiently. Learn its strengths and weaknesses.

You may use your calculator any way you wish. There are four types of things you should definitely know how to do. They are:

- Plot the graph of a function within an arbitrary viewing window,

- Find the zeros of functions (solve equations numerically),

- Calculate the numerical value of a derivative at a point, and

- Calculate the numerical value of a definite integral.

You may have programs in your calculator, but you will not be asked to use them. The questions on the exam are designed so that someone with a program, or a more expensive calculator, has no advantage over someone who does not. This includes many of the built-in programs.

Be sure your calculator is set in Radian mode.

Numerical answers may be left unsimplified and in terms of π, e, etc. There is no reason to change an answer to a decimal if you don't have to. (Why take the chance of pushing the wrong button?)

Install fresh batteries before the exam.

The Format of the Exams

There are two parts to the AP Exams: a multiple-choice section and a free-response section. The number of questions and timing may change slightly from year to year. Be sure you check the current College Board publications for your exam.

Both sections count equally towards your final grade. Both sections cover the full range of topics. It is natural to expect that different classes will cover some topics in greater detail than others; the exam will evaluate your knowledge of the calculus. It is not necessary to answer all the questions to get a good score. In fact, the exam is made so that the average score will be about 50%, this is usually a score of 3. You are probably used to class averages much higher than 50%; this test is different. Expect not to be able to answer some questions, and don't worry about it. Use your time on the ones you can answer.

The Current AP Calculus Exam Format is

Section I Part A (55 minutes) 28 multiple-choice questions for which you may not use a calculator.

Section I Part B (50 minutes) 17 multiple-choice questions. You may use your calculator on this section. Some of these questions require the use of a graphing calculator, others do not.

Section II Part A (45 minutes) Three free-response questions. You may use your calculator on this section. In this section you will find longer questions with several related parts. You are required to show your work in this section. You may continue work on this section without a calculator after you start part B.

Section II Part B (45 minutes) Three free-response questions. You may not use your calculator on this section. In this section you will find longer questions with several related parts. You are required to show your work in this section. You may use part of this time to work on Section II, Part A without a calculator

Multiple Choice Questions

Read each question carefully and look at the answer choices. Do the ones you are sure of. Don't struggle over one that isn't working out. Remember your time is limited and you do not need to answer all of the questions. There is a penalty for guessing, so don't guess blindly. You receive one point for each correct answer. One-quarter point is deducted for each wrong answer. Nothing is deducted for a question that is left blank. Guessing may improve your score only if you can eliminate one or more of the choices. Be sure to bubble your answer in the correct space on the answer sheet.

Types of Multiple-choice Questions

- One type of question may ask for a computation (a limit, a derivative, a definite or indefinite integral) and give five possible answers. Be aware that answers which result from predictable mistakes are among the choices — work carefully, just because your answer is there doesn't mean it's correct.

- Another type may ask you only to set up a problem. Looking at the answer choices may keep you from doing too much work.

- Some questions ask you to choose the one true or one false statement from a list of five statements. Be sure you know if you are looking for a true or a false statement.

- Another type of question asks which of three statements is true (or false). The answer may be any one or some combination of the statements.

- Another type may ask you to choose the correct table or graph from among five choices.

Free-response Questions

The general directions for Section II require you to show your work and indicate the methods you use to arrive at your answers. In addition, parts of questions may say, "Justify your answer" or "Show the analysis that leads to your conclusion." Your answers will be read by calculus teachers who will judge your work. It is important that you clearly show how you arrived at your answer. Unsupported answers lose points even if the final answer is correct.

The questions are designed to show the breadth and depth of your knowledge. There are some common types of questions that are asked. There will also be questions asked in new and original ways.

Some Things to Keep in Mind About Free-response Questions:

- Don't write a long essay. It's not necessary. Do show the work that you do so that the reader will understand you. You may use common terms and names like "the First Derivative Test." You do not need to name theorems. A number line analysis of a first or second derivative is not sufficient justification; you must include a brief written statement, which may be based on your number line, explaining how you know your answer is true.

- The free-response section of the exam rarely requires long complicated computation. If you find yourself doing a long complicated computation, you've probably gone wrong somewhere and should start over.

- Expect to be asked to explain or interpret something. Answer with one or two short sentences. It is one way you will be tested on the "verbal" part of the Rule of Four.

- Do not explain how to do the problem you cannot do. A general explanation without work will receive no credit. You must do the problem you are given.

- Avoid simplifying numerical answers. Answers may be left unsimplified as fractions, radicals, powers of e, in terms of π, etc. Do not take a chance of pushing the wrong button once you have an acceptable answer. If you do arithmetic, it must be done correctly. Every year students find the correct answer, change it to a decimal incorrectly, and lose a point. Decimal answers (for example a definite integral on a calculator) are acceptable even if an exact answer is possible.

- If you make a mistake, cross it out. Crossed out work is not read or graded. If you leave wrong work on your paper, not crossed out, it will be read and may affect your score.

- If you work the problem two different ways, choose the best one and put an X through the other. If both are left, they will both be scored and the scores will be averaged. This can lower your score even if one solution is perfect.

- Standard notation must be used. Don't use calculator notation. For example: fnInt(x^2,x,0,2) is not acceptable, use the standard $\int_0^2 x^2 dx$.

- Answers without work do not receive full credit. Don't do work on a calculator without indicating what you are doing. For example, if you are evaluating a definite integral, write the integral on your paper and put the calculator answer next to it. You do not need to show the work in-between (the antiderivative).

- Different calculators have different built-in utilities (for example, the ability to find points of inflection, or maximum values of a function). You may have programs in your calculator to do things such as the Trapezoidal Rule. However, if you use such a built-in utility, or a special program to do something other than the four things listed previously, you must show the complete setup (the terms of the Trapezoidal Rule, the computation and analysis of the second derivative required to find a point of inflection etc.) on your paper. Only the four things listed may be done without further explanation.

- Don't put things where they are not needed. Work must be shown on the part of the answer booklet where it is used. For example, if you need a derivative in part (b) of a question and you have it in part (a), where it is not needed, you will not get credit for finding the derivative (in either part). Either copy it in part (b), or draw an arrow over to where you wrote it. You must show that you know where you need the derivative as well as your ability to find it. Likewise, do not put work on a graphs, table of values, or drawing. Do not put work on the questions sheets and do not put work on the graphs or drawing on the questions sheets. These pages are not returned with your answers and the readers do not see them..

- Finally, the parts of a free-response question are related to each other. This can help you in two ways:

  o Sometimes each part may be answered without reference to the other parts. Read and try all of the parts. If you cannot do part (a), maybe you can do part (b). Perhaps doing part (b) will give you a hint on how to do part (a).

  o Other times the one part will lead to the next. This is done to help you find your way through the problem. Keep in mind that this may be the case, and work your way from part (a), to part (b), to part (c), even if you're not sure where the problem is heading.

- Try all of the free-response questions. They are written so that the first parts are easier in order to help you get started. Even if you don't get the entire problem, some points are better than no points.

Common Free-response Mistakes

- Algebra and arithmetic mistakes.

- Missing limits of integration.

- Not considering the end points of an interval (for example, when looking for the absolute maximum value of a function).

- Giving answers from points outside the given interval.

- Not giving both coordinates of a point when required.

- Giving both coordinates when only one is asked for; remember, "value of a function" means the y-value.

- Having your calculator in degrees mode.

- Not answering the question that was asked even though all the work is correct. If it is a yes or no question, say "yes" or "no."

- Ignoring units of measure.

- Family of function problems: Questions that start with a phrase like, "This question deals with functions defined by $f(x) = 1 + b\sin(x)$ where b is a positive constant..." are meant to be done in general, not for a specific value of b. Even if you get the correct answer using a specific value of b, you may lose points. The reason is that, because you used a particular value, you have no way to be sure that your answers are true for all values of b.

- Don't Curve Fit: Occasionally, a function is given as a graph or a table of values with no equation. You are being asked to demonstrate that you can work from the graphical or numerical data. The questions that follow can be answered without an equation. You may have learned to approximate functions using various curve fitting (regression) operations built into your calculator. This should be avoided. While this is a perfectly good approach in the real world, you may lose points because you are not working with the function you were given, (only an approximation of it), and curve fitting is not one of the four allowed calculator operations.

- Using a built-in calculator utility or a program without showing all the work and justification for what you are doing. You may do only the four things listed on page ii without further explanation.

A Word About Three-Decimal Place Accuracy.

Some answers, the evaluation of definite integrals is a prime example, must be written as decimals because they are found using a graphing calculator. These answers, and other answers that you choose to change to decimals, must be correct to three places past the decimal point. This means that the answer may be rounded to three decimal places, truncated after the third decimal place, or left with more than three decimal places, as long as the first three are correct. An answer of π, which should be left as π, may be given as 3.1415926535898..., 3.142, 3.141, or even 3.14199999. If the number ends in zeros, they may be omitted; thus, 17.320 may be given as 17.32, and 56.000 may be given as 56.

Too often students may choose to give decimal answers when they are not required. Once a free-response answer is entirely in terms of numbers, there is no need to change the number to a decimal. For example, 1999 AB 1(c) does not require a decimal answer: $-\frac{1}{2}\cos 4 + \frac{7}{2}$ is sufficient. If the decimal is correct (to three decimal places) you will receive the credit. However, if you change a correct answer to an incorrect decimal (including one with too few decimals) then you will lose credit. The moral is: avoid arithmetic, avoid decimals; give them only if you cannot give anything else.

Rounding too soon is another common mistake made by students. Computations should be done with more decimal places than is required in the final answer. Learn how to store the intermediate values in your calculator, and recall them when you need them in a computation. If premature rounding affects the three decimal place accuracy of the final answer, you will not be given the answer point. However, a rounded answer used in the next part of a problem will not be held against you.

Good Luck!

Sample Examination I

Directions: Solve each of the following problems, using the available space for scratchwork. After examining the form of the choices, decide which is the best of the choices given. Do not spend too much time on any one problem. Calculators may NOT be used on this part of the exam.

In this test: Unless otherwise specified, the domain of a function f is assumed to be the set of all real numbers x for which $f(x)$ is a real number.

1. If f is a continuous function defined by $f(x) = \begin{cases} x^2 + bx, & x \le 5 \\ 5\sin\left(\dfrac{\pi}{2}x\right), & x > 5 \end{cases}$

   then $b =$

   (A) -6

   (B) -5

   (C) -4

   (D) 4

   (E) 5

   Answer

2. The graph of $y = 3x^2 - x^3$ has a relative maximum at

   (A) $(0,0)$ only

   (B) $(1,2)$ only

   (C) $(2,4)$ only

   (D) $(4,-16)$ only

   (E) $(0,0)$ and $(2,4)$

   Answer

3. A particle moves in the xy-plane so that its velocity vector at time t is $v(t) = (t^2, \sin(\pi t))$ and the particle's position vector at time $t = 0$ is $(1, 0)$. What is the position vector of the particle when $t = 3$?

(A) $(9, \frac{1}{\pi})$

(B) $(10, \frac{2}{\pi})$

(C) $(6, -2\pi)$

(D) $(10, 2\pi)$

(E) $(10, 2)$

Answer

4. For what values of x does the curve $y^2 - x^3 - 15x^2 = 8$ have horizontal tangent lines?

(A) $x = -10$ only

(B) $x = 0$ only

(C) $x = 10$ only

(D) $x = 0$ and $x = -10$

(E) $x = -10$, $x = 0$, and $x = 10$

Answer

5. $\displaystyle\lim_{x\to\infty} \frac{10^8 x^5 + 10^6 x^4 + 10^4 x^2}{10^9 x^6 + 10^7 x^5 + 10^5 x^3} =$

(A) 0   (B) 1   (C) -1 (D) $\frac{1}{10}$ (E) $-\frac{1}{10}$

Answer

6. $\displaystyle\lim_{x\to 0} \frac{\int_1^{1+x} \frac{\cos t}{t}\, dt}{x}$ is

(A) $-\cos 1$ (B) $\cos 1$ (C) $-\sin 1$ (D) $\sin 1$ (E) nonexistent

Answer

7. If $f(x) = \sqrt{4\sin x + 2}$, then $f'(0) =$

(A) -2

(B) 0

(C) 1

(D) $\dfrac{\sqrt{2}}{2}$

(E) $\sqrt{2}$

Answer

8. If t is measured in hours and $f'(t)$ is measured in knots, then $\displaystyle\int_0^2 f'(t)\,dt =$ (Note: 1 knot = 1 nautical mile/hour)

   (A) $f(2)$ knots

   (B) $f(2) - f(0)$ knots

   (C) $f(2)$ nautical miles

   (D) $f(2) - f(0)$ nautical miles

   (E) $f(2) - f(0)$ knots/ hour

Answer

9. An equation of the tangent line to the curve $x^2 + y^2 = 169$ at the point $(5, -12)$ is

   (A) $5y - 12x = -120$

   (B) $5x - 12y = 119$

   (C) $5x - 12y = 169$

   (D) $12x + 5y = 0$

   (E) $12x + 5y = 169$

Answer

10.

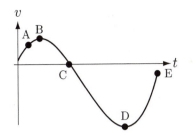

The figure above shows the graph of the velocity of a moving object as a function of time. At which of the marked points is the *speed* the greatest?

(A) A

(B) B

(C) C

(D) D

(E) E

Answer

11. What are all values of x for which the series $\displaystyle\sum_{n=2}^{\infty} \frac{(-1)^n}{\ln n} x^n$ converges?

(A) $-e < x \le e$

(B) $-1 \le x < 1$

(C) $-e \le x < e$

(D) $-1 < x \le 1$

(E) $-1 \le x \le 1$

Answer

12. $\int \dfrac{1}{\sqrt{4-x^2}}\, dx =$

(A) $\text{Arcsin}\left(\dfrac{x}{2}\right) + C$

(B) $2\sqrt{4-x^2} + C$

(C) $\text{Arcsin}\,(x) + C$

(D) $\sqrt{4-x^2} + C$

(E) $\dfrac{1}{2}\text{Arcsin}\left(\dfrac{x}{2}\right) + C$

Answer

13. If the graph of $f(x) = 2x^2 + \dfrac{k}{x}$ has a point of inflection at $x = -1$, then the value of k is

(A)   2

(B)   1

(C)   0

(D) −1

(E) −2

Answer

14. If $\int x\sec^2 x\, dx = f(x) + \ln|\cos x| + C$, then $f(x) =$

(A) $\tan x$ (B) $\dfrac{1}{2}x^2$ (C) $x\tan x$ (D) $x^2\tan x$ (E) $\tan^2 x$

Answer

15. Which of the following is an equation of the line tangent to the curve with parametric equations $x = 3e^{-t}$, $y = 6e^t$ at the point where $t = 0$?

(A) $2x + y - 12 = 0$

(B) $-2x + y - 12 = 0$

(C) $x - 2y + 9 = 0$

(D) $2x - y = 0$

(E) $x + 2y - 15 = 0$

Answer

16. $\displaystyle\int \frac{dx}{2x^2 + 3x + 1} =$

(A) $2\ln\left|\dfrac{2x + 1}{x + 1}\right| + C$

(B) $\ln\left|\dfrac{(2x + 1)^2}{x + 1}\right| + C$

(C) $\ln\left|\dfrac{x + 1}{2x + 1}\right| + C$

(D) $\ln\left|\dfrac{2x + 1}{x + 1}\right| + C$

(E) $\ln\left|(x + 1)(2x + 1)\right| + C$

Answer

17. If $x = \sin t$ and $y = \cos^2 t$, then $\dfrac{d^2y}{dx^2}$ at $t = \pi$ is

(A) -2 (B) $-\frac{1}{4}$ (C) 0 (D) $\frac{1}{4}$ (E) 2

Answer

18. If $y = x(\ln x)^2$, then $\dfrac{dy}{dx} =$

(A) $3(\ln x)^2$

(B) $(\ln x)(2x + \ln x)$

(C) $(\ln x)(2 + \ln x)$

(D) $(\ln x)(2 + x \ln x)$

(E) $(\ln x)(1 + \ln x)$

Answer

19. If $\displaystyle\int_0^6 (x^2 - 2x + 2)\,dx$ is approximated by a lower sum using three inscribed rectangles of equal width on the x–axis, then the approximation is

(A) 24   (B) 26   (C) 28

(D) 48   (E) 76

Answer

20.

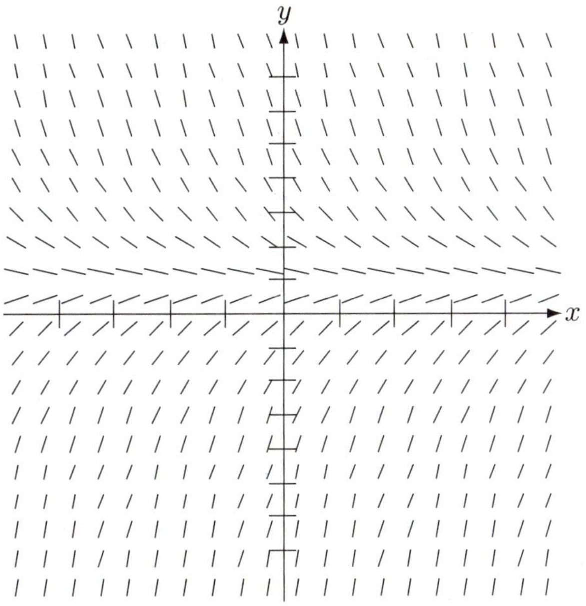

Shown above is the slope field for which differential equation?

(A) $\dfrac{dy}{dx} = 1 - x$

(B) $\dfrac{dy}{dx} = x - y$

(C) $\dfrac{dy}{dx} = -\dfrac{x}{y}$

(D) $\dfrac{dy}{dx} = 1 + y^2$

(E) $\dfrac{dy}{dx} = 1 - y$

Answer

21. The power series $1 + 2x + 4x^2 + 8x^3 + \cdots + 2^{n-1}x^{n-1} + \cdots$ converges for what values of x?

(A) $x = 0$ only   (B) $-\frac{1}{2} < x < \frac{1}{2}$ only   (C) $-1 < x < 1$ only

(D) $-2 < x < 2$ only   (E)   All real numbers x

Answer

22. The Taylor series for $\dfrac{\sin(x^2)}{x^2}$ centered at $x = 0$ is

(A) $\displaystyle\sum_{k=0}^{\infty} \frac{(-1)^k x^{2k+1}}{(2k+1)!}$

(B) $\displaystyle\sum_{k=0}^{\infty} \frac{(-1)^k x^{2k}}{(2k+1)!}$

(C) $\displaystyle\sum_{k=0}^{\infty} \frac{(-1)^k x^{2k+1}}{(2k)!}$

(D) $\displaystyle\sum_{k=0}^{\infty} \frac{(-1)^k x^{4k}}{(2k+1)!}$

(E) $\dfrac{1}{x} + \displaystyle\sum_{k=1}^{\infty} \frac{(-1)^k x^{2k-1}}{(2k-1)!}$

Answer

23. If the length of a curve $y = f(x)$ from $x = a$ to $x = b$ is given by $L = \displaystyle\int_a^b \sqrt{e^{2x} + 2e^x + 2} \; dx$, then $f(x)$ is

(A) $2e^{2x} + 2e^x$

(B) $\frac{1}{2}e^{2x} + 2e^x + 2x$

(C) $e^x - x + 3$

(D) $e^x + 1$

(E) $e^x + x - 2$

Answer

24.

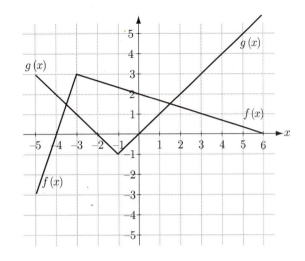

The functions f and g are piecewise linear functions whose graphs are shown above. If $h(x) = f(x)g(x)$, then $h'(3) =$

(A) $-\dfrac{8}{3}$　　　　　　(B) $-\dfrac{1}{3}$　　　　　　(C) 0

(D) $\dfrac{2}{3}$　　　　　　(E) $\dfrac{8}{3}$

Answer

25. For which pair of functions $f(x)$ and $g(x)$ below, will the $\lim\limits_{x \to \infty} \dfrac{f(x)}{g(x)} = 0$?

	$f(x)$	$g(x)$
(A)	e^x	x^2
(B)	e^x	$\ln x$
(C)	$\ln x$	e^x
(D)	x	$\ln x$
(E)	3^x	2^x

Answer

26. $\displaystyle\int_1^4 \frac{t-2}{(t+1)(t-4)}\,dt$ is found by using which of the following limits?

(A) $\displaystyle\lim_{b\to 2}\int_b^4 \frac{t-2}{(t+1)(t-4)}\,dt$

(B) $\displaystyle\lim_{b\to 1+}\int_b^4 \frac{t-2}{(t+1)(t-4)}\,dt$

(C) $\displaystyle\lim_{b\to 4-}\int_b^4 \frac{t-2}{(t+1)(t-4)}\,dt$

(D) $\displaystyle\lim_{b\to 1}\int_b^4 \frac{t-2}{(t+1)(t-4)}\,dt$

(E) $\displaystyle\lim_{b\to 4-}\int_1^b \frac{t-2}{(t+1)(t-4)}\,dt$

Answer

27. The average value of the function $f(x) = \cos\left(\frac{1}{2}x\right)$ on the closed interval $[-4, 0]$ is

(A) $-\frac{1}{2}\sin(2)$

(B) $-\frac{1}{4}\sin(2)$

(C) $\frac{1}{2}\cos(2)$

(D) $\frac{1}{4}\sin(2)$

(E) $\frac{1}{2}\sin(2)$

Answer

28. If n is a positive integer, then $\displaystyle\lim_{n\to\infty} \frac{1}{n}\left[\frac{1}{1+(1/n)} + \frac{1}{1+(2/n)} + \cdots + \frac{1}{1+(n/n)}\right]$ can be expressed as

(A) $\displaystyle\int_1^2 \frac{1}{x}\,dx$

(B) $\displaystyle\int_1^2 \frac{1}{x+1}\,dx$

(C) $\displaystyle\int_1^2 x\,dx$

(D) $\displaystyle\int_1^2 \frac{2}{x+1}\,dx$

(E) $\displaystyle\int_0^1 \frac{1}{x}\,dx$

Answer

<u>Section I Part B</u>

<u>Directions</u>: Solve each of the following problems, using the available space for scratchwork. After examining the form of the choices, decide which is the best of the choices given. Do not spend too much time on any one problem. A graphing calculator is required for some questions on this part of the examination.

<u>In this test</u>:

(1)   The exact numerical value of the correct answer does not always appear among the choices given. When this happens, select from among the choices the number that best approximates the exact numerical value.

(2)   Unless otherwise specified, the domain of a function f is assumed to be the set of all real numbers x for which $f(x)$ is a real number.

29. The volume of the solid formed by revolving the region bounded by the graph of $y = (x-3)^2$ and the coordinate axes about the <u>x-axis</u> is given by which of the following integrals?

(A) $\pi \displaystyle\int_0^3 (x-3)^2 \, dx$

(B) $\pi \displaystyle\int_0^3 (x-3)^4 \, dx$

(C) $\pi \displaystyle\int_0^3 (x-3)^3 \, dx$

(D) $\pi \displaystyle\int_0^3 x(x-3)^2 \, dx$

(E) $\pi \displaystyle\int_0^3 x(x-3)^4 \, dx$

Answer

30. Let f be the function given by $f(x) = \tan x$ and let g be the function given by $g(x) = x^2$. At what value of x in the interval $0 \le x \le \pi$ do the graphs of f and g have parallel tangent lines?

(A)   0   (B)   0.660   (C)   2.083

(D)   2.194   (E)   2.207

Answer

31. Let $f(t) = \frac{1}{t}$ for $t > 0$. For what value of t is $f'(t)$ equal to the average rate of change of f on the closed interval $[a, b]$?

(A) $-\sqrt{ab}$

(B) $\sqrt{ab}$

(C) $-\dfrac{1}{\sqrt{ab}}$

(D) $\dfrac{1}{\sqrt{ab}}$

(E) $\sqrt{\dfrac{1}{2}\left(\dfrac{1}{b} - \dfrac{1}{a}\right)}$

Answer

32. Let f be the function given by $f(x) = 3 + \displaystyle\int_0^x \cos(t^2)\,dt$. What is the smallest positive number a, for which $f'(a) = 0$?

(A) 1

(B) 1.253

(C) 1.571

(D) 1.772

(E) 3.142

Answer

33. Let $R(t)$ represent the rate in gallons/hour at which water is leaking out of a tank, where t is measured in hours. Which of the following expressions represents the total gallons of water that leaks out in the first three hours?

(A) $R(3) - R(0)$

(B) $\dfrac{R(3) - R(0)}{3 - 0}$

(C) $\displaystyle\int_0^3 R(t)\ dt$

(D) $\displaystyle\int_0^3 R'(t)\ dt$

(E) $\dfrac{1}{3} \displaystyle\int_0^3 R(t)\ dt$

Answer

34. Which of the following gives the area of the region enclosed by the graph of the polar curve $r = 1 + \cos\theta$?

(A) $\displaystyle\int_0^\pi (1 + \cos^2\theta)\ d\theta$

(B) $\displaystyle\int_0^\pi (1 + \cos\theta)^2\ d\theta$

(C) $\displaystyle\int_0^{2\pi} (1 + \cos\theta)\ d\theta$

(D) $\displaystyle\int_0^{2\pi} (1 + \cos\theta)^2\ d\theta$

(E) $\dfrac{1}{2} \displaystyle\int_0^{2\pi} (1 + \cos^2\theta)\ d\theta$

Answer

35.

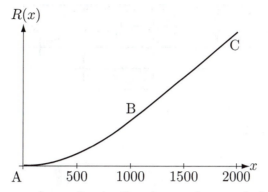

The figure above shows a road running in the shape of a parabola from the bottom of a hill at A to point B. At B it changes to a line and continues on to C. The equation of the road is

$$R(x) = \begin{cases} ax^2, & \text{from A to B} \\ bx + c, & \text{from B to C} \end{cases}$$

B is 1000 feet horizontally from A and 100 feet higher. Since the road is smooth, $R'(x)$ is continuous. What is the value of b?

(A) 0.2

(B) 0.02

(C) 0.002

(D) 0.0002

(E) 0.00002

Answer

36. $\displaystyle\sum_{k=0}^{\infty} \left(\sin\left(\frac{\pi}{6}\right)\right)^k$ is

(A) 1   (B) 2   (C) $\dfrac{1}{1 - \frac{\sqrt{3}}{2}}$ (D) $\dfrac{\frac{\sqrt{3}}{2}}{1 - \frac{\sqrt{3}}{2}}$ (E) divergent

Answer

37.

x	−0.3	−0.2	−0.1	0	0.1	0.2	0.3
$f(x)$	2.018	2.008	2.002	2	2.002	2.008	2.018
$g(x)$	1	1	1	2	2	2	2
$h(x)$	1.971	1.987	1.997	undefined	1.997	1.987	1.971

The table above gives the values of three functions, f, g, and h near $x = 0$. Based on the values given, for which function does it appear that the limit as x approaches zero is 2?

(A) f only

(B) g only

(C) h only

(D) f and h only

(E) f, g, and h

Answer

38. If $\dfrac{dy}{dx} = \dfrac{x^2}{y}$ and $x = 1$ when $y = 1$, then $y =$

(A) $\dfrac{2}{3}x^2 + 12$

(B) $-\sqrt{\dfrac{x^3 + 3}{3}}$

(C) $\sqrt{\dfrac{x^3 + 3}{3}}$

(D) $-\sqrt{\dfrac{2x^3 + 1}{3}}$

(E) $\sqrt{\dfrac{2x^3 + 1}{3}}$

Answer

39.

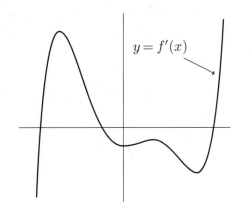

The figure above shows the graph of the derivative of a function f. How many points of inflection does f have in the interval shown?

(A) None   (B) One   (C) Two

(D) Three   (E) Four

Answer

40.

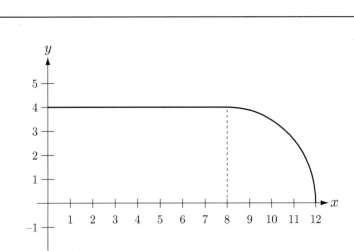

As shown in the figure above the function $f(x)$ consists of a line segment from $(0, 4)$ to $(8, 4)$ and one-quarter of a circle with a radius of 4. What is the average (mean) value of this function on the interval $[0, 12]$?

(A) 2   (B) 3.714   (C) 3.9

(D) 22.283   (E) 41.144

Answer

41. The graph of the third-degree Maclaurin polynomial that approximates $\sin(x)$ intersects the graph of $y = x^2$ at how many points?

(A) None

(B) One

(C) Two

(D) Three

(E) Four

Answer

42. The amount, $A(t)$, of a certain item produced in a factory is given by

$$A(t) = 4000 + 48(t - 3) - 4(t - 3)^3$$

where t is the number of hours of production since the beginning of the workday at 8:00 a.m. At what time is the rate of production increasing most rapidly?

(A) 8:00 a.m.

(B) 10:00 a.m.

(C) 11:00 a.m.

(D) 12:00 noon

(E) 1:00 p.m.

Answer

43. The derivative of $4x^2 \cos(x)$ is

    (A) $-8x \sin(x)$

    (B) $8x \cos(x) - 4x^2 \sin(x)$

    (C) $8x \cos(x) + 4x^2 \sin(x)$

    (D) $-8x \cos(x) - 4x^2 \sin(x)$

    (E) $-8x \cos(x) + 4x^2 \sin(x)$

Answer

44. A population grows according to the equation $P(t) = 6000 - 5500e^{-0.159t}$ for $t \geq 0$, t measured in years. This population will approach a limiting value as time goes on. During which year will the population reach <u>half</u> of this limiting value?

    (A) Second

    (B) Third

    (C) Fourth

    (D) Eighth

    (E) Twenty-ninth

Answer

45.

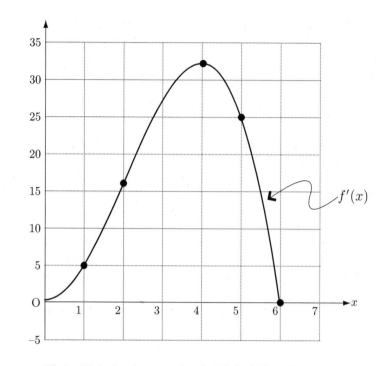

Note: This is the graph of $f'(x)$, NOT the
graph of $f(x)$.

Let f be a function defined for $0 \le x \le 6$. The graph of $f'(x)$ is shown above. If $f(2) = 10$,
which of the following best approximates the maximum value of $f(x)$?

(A) 30

(B) 50

(C) 70

(D) 90

(E) 110

Answer

SECTION II - FREE-RESPONSE QUESTIONS
GENERAL INSTRUCTIONS

You may wish to look over the problems before starting to work on them, since it is not expected that everyone will be able to complete all parts of all problems. All problems are given equal weight, but the parts of a particular problem are not necessarily given equal weight.

- YOU SHOULD WRITE ALL WORK FOR EACH PART OF EACH PROBLEM IN THE SPACE PROVIDED FOR THAT PART. Be sure to write clearly and legibly. If you make an error, you may save time by crossing it out rather than trying to erase it. Erased or crossed-out work will not be graded.

- Show all your work. Clearly label any functions, graphs, tables, or other objects that you use. You will be graded on the correctness and completeness of your methods as well as your answers. Answers without supporting work may not receive credit.

- Justifications require that you give mathematical (noncalculator) reasons.

- Your work must be expressed in standard mathematical notation rather than calculator syntax. For example, $\int_1^5 x^2\, dx$ may not be written as fnInt(X^2, X, 1, 5).

- Unless otherwise specified, answers (numeric or algebraic) need not be simplified.

- If you use decimal approximations in calculations, you will be graded on accuracy. Unless otherwise specified, your final answers should be accurate to three places after the decimal point.

- Unless otherwise specified, the domain of a function f is assumed to be the set of all real numbers x for which $f(x)$ is a real number.

SECTION II PART A: 45 Minutes, Questions 1,2,3

During the timed portion for Part A, you may work only on the problems in Part A. Write your solution to each part of each problem in the space provided for that part.

On Part A, you are permitted to use your calculator to solve an equation, find the derivative of a function at a point, or calculate the value of a definite integral. However, you must clearly indicate the setup of your problem, namely the equation, function, or integral you are using. If you use other built-in features or programs, you must show the mathematical steps necessary to produce your results.

Do not go on to Part B until you are told to do so.

SECTION II PART B: 45 Minutes, Questions 4,5,6

Write your solution to each part of each problem in the space provided for that part. During the timed portion for Part B, you may continue to work on the problems in Part A without the use of any calculator.

Section II Part A: Graphing Calculator MAY BE USED.

1.

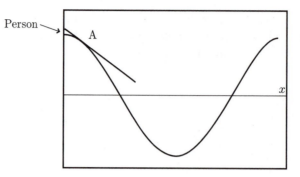

As shown in the figure above a person whose eye level is 5 feet above the ground stands on the top of a hill overlooking a valley. The shape of the valley is modeled by the graph of $f(x) = 50 \cos\left(\dfrac{x}{100}\right)$. The person's line of sight is tangent to the side of the hill at point A $\left(a, 50\cos\left(\dfrac{a}{100}\right)\right)$.

(a) Write an equation of the tangent line in terms of a.

(b) Find the value of a.

(c) Can the person see the top of a 25-foot tall house located at the lowest point of the valley? Justify your answer.

_____

(a) Write an equation of the tangent line in terms of a.

(b)   Find the value of a.

(c)   Can the person see the top of a 25-foot tall house located at the lowest point of the valley? Justify your answer.

Section II Part A: Graphing Calculator MAY BE USED.

2. Two particles move in the xy-plane. For time $0 \leq t \leq 2\pi$, the position of particle A is given by $x(t) = \cos t$ and $y(t) = t$, and the position of particle B is given by $x(t) = \sin t$ and $y(t) = t$.

   (a) In the viewing window provided below, sketch the path of particles A and B. Label the paths A and B and indicate with arrows the direction of each particle along its path.

   (b) Find the velocity vector for each particle.

   (c) At $t = 5$, which particle is moving to the right the fastest? Justify your answer.

   (d) At $t = 5$, which particle is moving upward the fastest? Justify your answer.

   (a) In the viewing window provided below, sketch the path of particles A and B. Label the paths A and B and indicate with arrows the direction of each particle along its path.

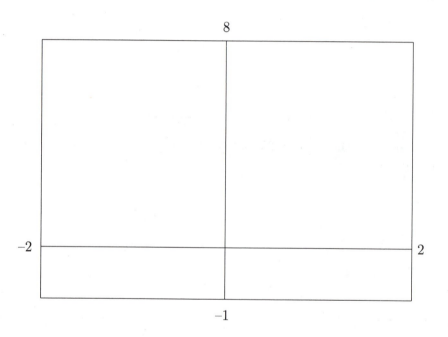

(b) Find the velocity vector for each particle.

(c) At $t = 5$, which particle is moving to the right the fastest? Justify your answer.

(d) At $t = 5$, which particle is moving upward the fastest? Justify your answer.

Section II Part A: Graphing Calculator MAY BE USED.

3. The number of minutes of daylight per day, $L(d)$, at 40°North latitude is modeled by the function

$$L(d) = 167.5 \sin \left(\frac{2\pi}{366} (d - 80) \right) + 731$$

where d is the number of days after the beginning of 2004. (For Jan. 1, 2004, $d = 1$; and for Dec. 31, 2004, $d = 366$, since 2004 was a leap year.)

(a)   Which day (d) has the most minutes of daylight? Justify your answer.

(b)   What is the average (mean) number of minutes of daylight in 2004? Justify your answer.

(c)   What is the total number of minutes of daylight in 2004? Justify your answer.

(a)   Which day (d) has the most minutes of daylight? Justify your answer.

(b)   What is the average (mean) number of minutes of daylight in 2004? Justify your answer.

(c)   What is the total number of minutes of daylight in 2004? Justify your answer.

4. Let R be the region to the right of $x = 1$ between the x-axis and the graph of $y = \dfrac{1}{x^2}$ and let S be the region to the right of $x = 1$ between horizontal line $y = 1$ and the graph of $y = \dfrac{1}{x^2}$.

    (a) Write an improper integral that gives the area of the region R and find the value (if any) to which it converges.

    (b) Write an improper integral that gives the area of the region S and find the value (if any) to which it converges.

    (c) The region R is the base of a solid. For this solid every cross section perpendicular to the x-axis is a square. Write an improper integral that gives the volume of this solid and find the value (if any) to which it converges.

    (a) Write an improper integral that gives the area of the region R and find the value (if any) to which it converges.

(b)   Write an improper integral that gives the area of the region S and find the value (if any) to which it converges.

(c)   The region R is the base of a solid. For this solid every cross section perpendicular to the x-axis is a square. Write an improper integral that gives the volume of this solid and find the value (if any) to which it converges.

Section II Part B: Graphing Calculator MAY **NOT** BE USED.

5. Consider the differential equation $\dfrac{dy}{dx} = xy^2$.

   (a) On the axes provided, sketch a slope field for the given differential equation at the nine points indicated.

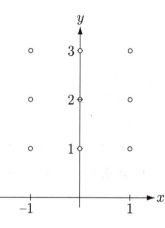

   (b) Find the general solution of the given differential equation in terms of a constant C.

   (c) Find the particular solution of the differential equation that satisfies the initial condition $y(0) = 1$.

   (d) For what values of the constant C will the solutions of the differential equation have one or more vertical asymptotes? Justify your answer.

   (a) On the axes provided, sketch a slope field for the given differential equation at the nine points indicated.

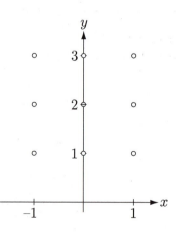

(b) Find the general solution of the given differential equation in terms of a constant C.

(c) Find the particular solution of the differential equation that satisfies the initial condition $y(0) = 1$.

(d) For what values of the constant C will the solutions of the differential equation have one or more vertical asymptotes? Justify your answer.

Section II Part B: Graphing Calculator MAY **NOT** BE USED.

6.

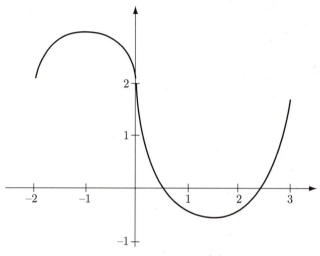

The graph of $f(x)$

Let f be a continuous function defined on the closed interval $[-2, 3]$. The graph of f consists of a semicircle and a semi-ellipse, as shown above. Let $G(x) = G(-2) + \int_{-2}^{x} f(t) \, dt$.

(a)   On what intervals, if any, is G concave down? Justify your answer.

(b)   If the equation of the line tangent to the graph of $G(x)$ at the point where $x = 0$ is $y = mx + 7$, what is the value of m and the value of $G(0)$? Justify your answer.

(c)   If the average value of f on the interval $0 \leq x \leq 3$ is zero, find the value of $G(3)$. Show the work that leads to your answer.

(a)   On what intervals, if any, is G concave down? Justify your answer.

(b)   If the equation of the line tangent to the graph of $G(x)$ at the point where $x = 0$ is $y = mx + 7$, what is the value of m and the value of $G(0)$? Justify your answer.

(c)   If the average value of f on the interval $0 \leq x \leq 3$ is zero, find the value of $G(3)$. Show the work that leads to your answer.

Sample Examination II

Directions: Solve each of the following problems, using the available space for scratchwork. After examining the form of the choices, decide which is the best of the choices given. Do not spend too much time on any one problem. Calculators may NOT be used on this part of the exam.

In this test: Unless otherwise specified, the domain of a function f is assumed to be the set of all real numbers x for which $f(x)$ is a real number.

1. $\displaystyle\int_0^2 (2x^3 + 3)\,dx =$

   (A) 8

   (B) 11

   (C) 14

   (D) 20

   (E) 24

Answer

2. In decomposing $\dfrac{5x - 2}{(x - 7)(x + 4)}$ by the method of partial fractions, one of the fractions obtained is

   (A) $\dfrac{-2}{x - 7}$ (B) $\dfrac{2}{x - 7}$ (C) $\dfrac{3}{x - 7}$ (D) $\dfrac{3}{x + 4}$ (E) $\dfrac{5}{x + 4}$

Answer

3. A particle moves in the xy-plane so that at any time t its coordinates are $x = t^2$ and $y = 4 - t^3$. At $t = 1$, its acceleration vector is

(A) $(2, -3)$ (B) $(2, -6)$ (C) $(1, -6)$ (D) $(2, 6)$ (E) $(1, -2)$

Answer

4. If $f(x) = (2 + 3x)^4$, then the fourth derivative of f is

(A) 0

(B) $4!(3)$

(C) $4!(3^4)$

(D) $4!(3^5)$

(E) $4!(2 + 3x)$

Answer

5. At what value(s) of x does $f(x) = x^4 - 8x^2$ have a relative minimum?

(A) 0 and -2 only

(B) 0 and 2 only

(C) 0 only

(D) -2 and 2 only

(E) -2, 0, and 2

Answer

6. $\displaystyle\int \sqrt{x}(x+2)\,dx =$

(A) $\dfrac{2}{5}x^{\frac{5}{2}} + \dfrac{4}{3}x^{\frac{3}{2}} + C$ (B) $\dfrac{2}{5}x^{\frac{3}{2}} + \dfrac{4}{3}x^{\frac{1}{2}} + C$ (C) $\dfrac{3}{2}\sqrt{x} + \dfrac{1}{\sqrt{x}} + C$

(D) $\sqrt{x^3} + 2\sqrt{x} + C$ (E) $\dfrac{x^2}{2}\left(\dfrac{2}{3}x^{\frac{3}{2}} + 2x^{\frac{1}{2}}\right) + C$

Answer

☐

---

7. A curve in the xy-plane is defined parametrically by the equations $x = t^2 + t$ and $y = t^2 - t$. For what value of t is the tangent line to the curve horizontal?

(A) $t = -1$

(B) $t = -\dfrac{1}{2}$

(C) $t = 0$

(D) $t = \dfrac{1}{2}$

(E) $t = 1$

Answer

☐

8. The function $y = x^4 + bx^2 + 8x + 1$ has a horizontal tangent and a point of inflection for the same value of x. What must be the value of b?

   (A) -6

   (B) -1

   (C)   1

   (D)   4

   (E)   6

Answer

9. Let $y = f(x)$ be the solution to the differential equation $\dfrac{dy}{dx} = y - x$. The point $(5, 1)$ is on the graph of the solution to this differential equation. What is the approximation for $f(6)$ if Euler's Method is used, starting at $x = 5$ with a step size of 0.5?

   (A) -4.25

   (B) -3.25

   (C) -1.25

   (D)   0.75

   (E)   2.25

Answer

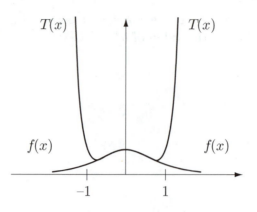

10. The figure above shows the graph of $y = f(x)$ and $y = T(x)$ where $T(x)$ is a Taylor polynomial for $f(x)$ centered at zero. Which of the following statements are true?

> I. $T(0.5)$ is a good approximation for $f(0.5)$.
>
> II. $T(1.5)$ is a good approximation for $f(1.5)$.
>
> III. $T(0) = f(0)$

(A) I only

(B) II only

(C) III only

(D) I and III only

(E) I, II, and III

Answer

11. $\lim\limits_{x \to 2} \dfrac{x^2 - 4}{\int_2^x \cos(\pi t)\, dt}$ is

(A) 0   (B) 1   (C) 2

(D) 4   (E) undefined

Answer

12. Which of the following improper integrals converge?

$$\text{I.} \quad \int_0^\infty e^{-x}\, dx$$

$$\text{II.} \quad \int_0^1 \frac{1}{x^2}\, dx$$

$$\text{III.} \quad \int_0^1 \frac{1}{\sqrt{x}}\, dx$$

(A) I only

(B) III only

(C) I and II only

(D) II and III only

(E) I and III only

Answer

13. If $x + y = xy$, then $\dfrac{dy}{dx}$ is

(A) $\dfrac{1}{x-1}$

(B) $\dfrac{1-y}{x-1}$

(C) $\dfrac{y-1}{x-1}$

(D) $x + y - 1$

(E) $\dfrac{2-xy}{y}$

Answer

14. If f and g are continuously differentiable functions defined for all real numbers, which of the following definite integrals is equal to $f(g(4)) - f(g(2))$?

(A) $\displaystyle\int_2^4 f'(g(x))\, dx$

(B) $\displaystyle\int_2^4 f(g(x))f'(x)\, dx$

(C) $\displaystyle\int_2^4 f(g(x))g'(x)\, dx$

(D) $\displaystyle\int_2^4 f(g'(x))g'(x)\, dx$

(E) $\displaystyle\int_2^4 f'(g(x))g'(x)\, dx$

Answer

15. The velocity of a particle moving along the y-axis is given by $v(t) = 8 - 2t$ for $t \geq 0$. The particle moves upward until it reaches the origin and then moves downward. The position of the particle at any time t is given by

(A) $-t^2 + 8t - 16$

(B) $-t^2 + 8t + 16$

(C) $2t^2 - 8t - 16$

(D) $8t - 2t^2$

(E) $8t - t^2$

Answer

16. If the substitution $u = \sqrt{x-1}$ is made, the integral $\int_2^5 \frac{\sqrt{x-1}}{x}\, dx =$

   (A) $\int_2^5 \frac{2u^2}{u^2+1}\, du$

   (B) $\int_1^2 \frac{u}{u^2+1}\, du$

   (C) $\int_1^2 \frac{u^2}{2(u^2+1)}\, du$

   (D) $\int_2^5 \frac{u}{u^2+1}\, du$

   (E) $\int_1^2 \frac{2u^2}{u^2+1}\, du$

Answer

17. The first three nonzero terms in the Maclaurin series about $x = 0$ of xe^{-x} are

   (A) $x - x^2 - \frac{x^3}{2!}$

   (B) $x - x^2 + \frac{x^3}{2!}$

   (C) $-x + x^2 - \frac{x^3}{2!}$

   (D) $x + x^2 + \frac{x^3}{2!}$

   (E) $1 - x + \frac{x^2}{2!}$

Answer

18. If $\int_0^2 (2x^3 - kx^2 + 2k)\, dx = 12$, then k must be

(A) -3

(B) -2

(C)   1

(D)   2

(E)   3

Answer

19. $\displaystyle\sum_{n=1}^{\infty} \left(\frac{1}{2}\right)^{2n}$ is

(A) $\dfrac{1}{3}$

(B) $\dfrac{1}{2}$

(C) 1

(D) 2

(E) ∞

Answer

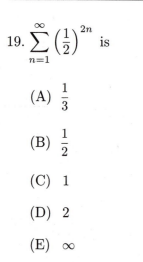

20. For $|x| < 1$, the derivative of $y = \ln \sqrt{1 - x^2}$ is

(A) $\dfrac{x}{1 - x^2}$ (B) $\dfrac{-x}{1 - x^2}$ (C) $\dfrac{-x^2}{x^2 - 1}$ (D) $\dfrac{1}{2(1 - x^2)}$ (E) $\dfrac{1}{\sqrt{1 - x^2}}$

Answer

21. A function whose derivative is a constant multiple of the function itself must be

(A) periodic

(B) linear

(C) logarithmic

(D) quadratic

(E) exponential

Answer

22. What are all values of x for which the graph of $y = x^3 - 6x^2$ is concave downward?

(A) $0 < x < 4$

(B) $x < 2$

(C) $x > 2$

(D) $x < 0$

(E) $x > 4$

Answer

23. The rate of decay of a radioactive substance is proportional to the amount of substance present. Four years ago there were 12 grams of the substance. Now there are 8 grams. How many grams will there be 8 years from now?

(A) 0

(B) $\dfrac{8}{3}$

(C) $\dfrac{32}{9}$

(D) $\dfrac{81}{16}$

(E) $\dfrac{16}{3}$

Answer

24. Which of the following series are convergent?

$$\text{I.} \qquad 1 + \frac{1}{2\sqrt{2}} + \frac{1}{3\sqrt{3}} + \cdots + \frac{1}{n\sqrt{n}} + \cdots$$

$$\text{II.} \qquad \frac{1}{1\cdot 2} + \frac{1}{2\cdot 3} + \frac{1}{3\cdot 4} + \cdots + \frac{1}{n(n+1)} + \cdots$$

$$\text{III.} \qquad 1 + \frac{1}{\ln 2} + \frac{1}{\ln 3} + \cdots + \frac{1}{\ln(n+1)} + \cdots$$

(A) I only

(B) II only

(C) I and II only

(D) II and III only

(E) I, II, and III

Answer

25. For all x if $f(x) = \displaystyle\sum_{n=0}^{\infty} \frac{(-1)^{n+1} x^{2n+1}}{(2n+1)!}$, then $f'(x) =$

(A) $\displaystyle\sum_{n=0}^{\infty} \frac{(-1)^{n+1} x^{2n}}{(2n+1)!}$

(B) $\displaystyle\sum_{n=0}^{\infty} \frac{(-1)^{n+1} x^{2n}}{(2n)!}$

(C) $\displaystyle\sum_{n=0}^{\infty} \frac{(-1)^{n+1} x^{2n}}{(2n+2)!}$

(D) $\displaystyle\sum_{n=0}^{\infty} \frac{(-1)^{n} x^{2n}}{(2n)!}$

(E) $\displaystyle\sum_{n=0}^{\infty} \frac{(-1)^{n} x^{2n}}{(2n+1)!}$

Answer

26. A normal line to the graph of a function f at the point $(x, f(x))$ is defined to be the line perpendicular to the tangent line at that point. An equation of the normal line to the curve $y = \sqrt[3]{x^2 - 1}$ at the point where $x = 3$ is

(A) $y + 12x = 38$

(B) $y - 4x = 10$

(C) $y + 2x = 4$

(D) $y + 2x = 8$

(E) $y - 2x = -4$

Answer

27. Which graph best represents the position of a particle, $s(t)$, as a function of time, if the particle's velocity and acceleration are both positive?

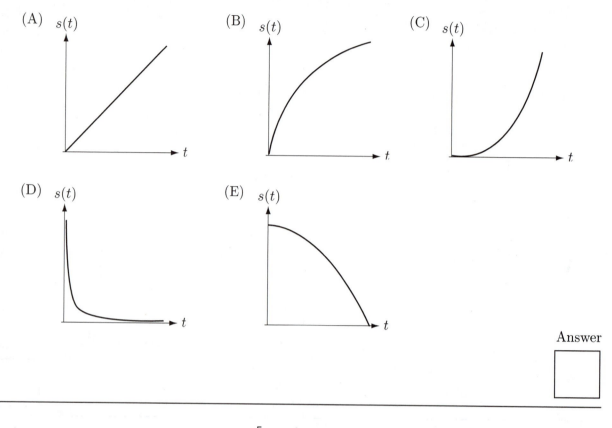

(A) $s(t)$

(B) $s(t)$

(C) $s(t)$

(D) $s(t)$

(E) $s(t)$

Answer

28. If n is a positive integer, then $\displaystyle\lim_{n\to\infty} \frac{1}{n}\left[\left(\frac{1}{n}\right)^2 + \left(\frac{2}{n}\right)^2 + \cdots + \left(\frac{n-1}{n}\right)^2\right] =$

(A) $\displaystyle\int_0^1 \frac{1}{x^2}\, dx$

(B) $\displaystyle\int_0^1 x^2\, dx$

(C) $\displaystyle\int_0^1 \frac{2}{x^2}\, dx$

(D) $\displaystyle\int_0^1 \frac{1}{x}\, dx$

(E) $\displaystyle\int_0^2 x^2\, dx$

Answer

Section I Part B

Directions: Solve each of the following problems, using the available space for scratchwork. After examining the form of the choices, decide which is the best of the choices given. Do not spend too much time on any one problem. A graphing calculator is required for some questions on this part of the examination.

In this test:

(1)   The exact numerical value of the correct answer does not always appear among the choices given. When this happens, select from among the choices the number that best approximates the exact numerical value.

(2)   Unless otherwise specified, the domain of a function f is assumed to be the set of all real numbers x for which $f(x)$ is a real number.

29. If the position of a particle moving in the xy-plane is given by the parametric equations $x(t) = 9\cos t$ and $y(t) = 4\sin t$ for $t \geq 0$, then at $t = 3$, the acceleration vector is

(A) $(-8.910, 0.564)$

(B) $(-1.270, -3.960)$

(C) $(8.910, -0.564)$

(D) $(8.910, 0.564)$

(E) $(-0.564, 8.910)$

Answer

30.

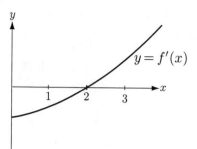

The graph of the derivative of a twice-differentiable function f is shown above. If $f(1) = -2$, which of the following is true?

(A) $f(2) < f'(2) < f''(2)$

(B) $f''(2) < f'(2) < f(2)$

(C) $f'(2) < f(2) < f''(2)$

(D) $f(2) < f''(2) < f'(2)$

(E) $f'(2) < f''(2) < f(2)$

Answer

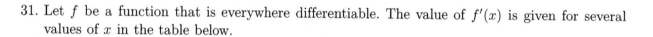

31. Let f be a function that is everywhere differentiable. The value of $f'(x)$ is given for several values of x in the table below.

x	−10	−5	0	5	10
$f'(x)$	−2	−1	0	1	2

If $f'(x)$ is always increasing, which statement about $f(x)$ must be true?

(A) $f(x)$ has a relative minimum at $x = 0$.

(B) $f(x)$ is concave downwards for all x.

(C) $f(x)$ has a point of inflection at $(0, f(0))$.

(D) $f(x)$ passes through the origin.

(E) $f(x)$ is an odd function.

Answer

32. A certain species of fish will grow from x million to $x(15 - x)$ million each year. In order to sustain a steady catch each year, a limit of $x(15 - x) - x$ million fish are to be caught, leaving x million fish to reproduce each year. What is the number of fish which should be left to reproduce each year so that the maximum catch may be sustained from year to year?

(A)   5 million

(B)   7 million

(C) 7.5 million

(D)   10 million

(E)   15 million

Answer

33.

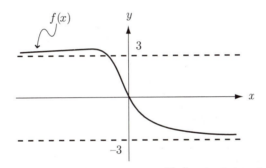

The figure above shows the graph of a function $f(x)$ which has horizontal asymptotes of $y = 3$ and $y = -3$. Which of the following statements are true?

$$\text{I.} \quad f'(x) < 0 \text{ for all } x \geq 0$$

$$\text{II.} \quad \lim_{x \to +\infty} f'(x) = 0$$

$$\text{III.} \quad \lim_{x \to -\infty} f'(x) = 3$$

(A) I only

(B) II only

(C) III only

(D) I and II only

(E) I, II, and III

Answer

34. The velocity vector of a particle moving in the coordinate plane is $(4t, -2t)$ for $t \geq 0$. The path of the particle lies on

(A) a hyperbola

(B) an ellipse

(C) a line

(D) a parabola

(E) a ray

Answer

35.

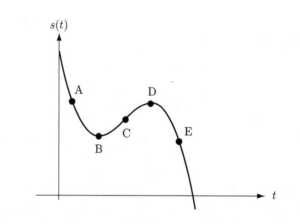

The graph above shows the distance $s(t)$ from a reference point of a particle moving on a number line, as a function of time. Which of the points marked is closest to the point where the acceleration first becomes negative?

(A) A

(B) B

(C) C

(D) D

(E) E

Answer

36. The <u>derivative</u> of f is given by $f'(x) = e^x(-x^3 + 3x) - 3$, for $0 \le x \le 5$.

At what value of x does $f(x)$ attain its absolute minimum?

(A) For no value of x

(B) 0

(C) 0.618

(D) 1.623

(E) 5

Answer

37.

x	$f(x)$
3.99800	1.15315
3.99900	1.15548
4.00000	1.15782
4.00100	1.16016
4.00200	1.16250

The table above gives values of a differentiable function f. What is the approximate value of $f'(4)$?

(A) 0.00234

(B) 0.289

(C) 0.427

(D) 2.340

(E) 4.270

Answer

38. If $y = 7$ is a horizontal asymptote of a rational function f, then which of the following must be true?

(A) $\lim\limits_{x \to 7} f(x) = \infty$

(B) $\lim\limits_{x \to \infty} f(x) = 7$

(C) $\lim\limits_{x \to 0} f(x) = 7$

(D) $\lim\limits_{x \to 7} f(x) = 0$

(E) $\lim\limits_{x \to -\infty} f(x) = -7$

Answer

39. In the interval $0 \le x \le 5$ the graphs of $y = \cos 2x$ and $y = \sin 3x$ intersect four times. Let A, B, C, and D be the x-coordinates of these points so that $0 < A < B < C < D < 5$. Which of the definite integrals below represents the largest number?

(A) $\displaystyle\int_0^A (\cos 2x - \sin 3x)\ dx$

(B) $\displaystyle\int_A^B (\sin 3x - \cos 2x)\ dx$

(C) $\displaystyle\int_B^C (\sin 3x - \cos 2x)\ dx$

(D) $\displaystyle\int_C^D (\sin 3x - \cos 2x)\ dx$

(E) $\displaystyle\int_C^D (\cos 2x - \sin 3x)\ dx$

Answer

40. The function $f(x) = \tan(3^x)$ has one zero in the interval $[0, 1.4]$. The derivative at this point is

(A) 0.411

(B) 1.042

(C) 3.451

(D) 3.763

(E) undefined

Answer

41.

x	0	1	2	3	4	5	6
$f(x)$	0	0.25	0.48	0.68	0.84	0.95	1

For the values of a continuous function given in the table above, $\int_0^6 f(x)\, dx$ is approximated by a Riemann sum using the value at the midpoint of each of three intervals of width 2. The approximation is

(A) 2.64

(B) 3.64

(C) 3.72

(D) 3.76

(E) 4.64

Answer

42. $\dfrac{d}{dx} \displaystyle\int_{x}^{x^3} \sin(t^2)\, dt =$

(A) $\sin(x^6) - \sin(x^2)$

(B) $6x^2 \sin(x^3) - 2\sin x$

(C) $3x^2 \sin(x^6) - \sin(x^2)$

(D) $6x^5 \sin(x^6) - 2\sin(x^2)$

(E) $2x^3 \cos(x^6) - 2x\cos(x^2)$

Answer

43. A tank is being filled with water at the rate of $300\sqrt{t}$ gallons per hour with $t > 0$ measured in hours. If the tank is originally empty, how many gallons of water are in the tank after 4 hours?

(A)   600

(B)   900

(C) 1200

(D) 1600

(E) 2400

Answer

44. The region in the first quadrant enclosed by the graphs of $y = x$ and $y = 2\sin x$ is revolved about the x-axis. The volume of the solid generated is

(A) 1.895

(B) 2.126

(C) 5.811

(D) 6.678

(E) 13.355

Answer

45. The tangent line to the graph $y = e^{2-x}$ at the point $(1, e)$ intersects both coordinate axes. What is the area of the triangle formed by this tangent line and the coordinate axes?

(A) $2e$

(B) $e^2 - 1$

(C) e^2

(D) $2e\sqrt{e}$

(E) $4e$

Answer

SECTION II - FREE-RESPONSE QUESTIONS
GENERAL INSTRUCTIONS

You may wish to look over the problems before starting to work on them, since it is not expected that everyone will be able to complete all parts of all problems. All problems are given equal weight, but the parts of a particular problem are not necessarily given equal weight.

- YOU SHOULD WRITE ALL WORK FOR EACH PART OF EACH PROBLEM IN THE SPACE PROVIDED FOR THAT PART. Be sure to write clearly and legibly. If you make an error, you may save time by crossing it out rather than trying to erase it. Erased or crossed-out work will not be graded.

- Show all your work. Clearly label any functions, graphs, tables, or other objects that you use. You will be graded on the correctness and completeness of your methods as well as your answers. Answers without supporting work may not receive credit.

- Justifications require that you give mathematical (noncalculator) reasons.

- Your work must be expressed in standard mathematical notation rather than calculator syntax. For example, $\int_1^5 x^2 \, dx$ may not be written as fnInt(X^2, X, 1, 5).

- Unless otherwise specified, answers (numeric or algebraic) need not be simplified.

- If you use decimal approximations in calculations, you will be graded on accuracy. Unless otherwise specified, your final answers should be accurate to three places after the decimal point.

- Unless otherwise specified, the domain of a function f is assumed to be the set of all real numbers x for which $f(x)$ is a real number.

SECTION II PART A: 45 Minutes, Questions 1,2,3

During the timed portion for Part A, you may work only on the problems in Part A. Write your solution to each part of each problem in the space provided for that part.

On Part A, you are permitted to use your calculator to solve an equation, find the derivative of a function at a point, or calculate the value of a definite integral. However, you must clearly indicate the setup of your problem, namely the equation, function, or integral you are using. If you use other built-in features or programs, you must show the mathematical steps necessary to produce your results.

Do not go on to Part B until you are told to do so.

SECTION II PART B: 45 Minutes, Questions 4,5,6

Write your solution to each part of each problem in the space provided for that part. During the timed portion for Part B, you may continue to work on the problems in Part A without the use of any calculator.

1. A train travels along a straight track. Its velocity, $v(t)$, in miles per hour for the first half of the trip is given by $v(t) = -5t^2 + 20t + 25$, $0 \leq t \leq 5$, t is time in hours. During the second half of the trip its velocity is given by the values in the table below. Note that $v(9) = A$.

t	5	6	7	8	9	10
$v(t)$	0	-25	-20	-50	A	-20

(a)   How far did the train travel in the first half of the trip? Include units of measure.

(b)   What was the acceleration at $t = 3$ hours? Include units of measure.

(c)   If the train has returned to its starting point at $t = 10$ hours, estimate the value of A to the nearest whole number. Show how you arrived at your answer.

(a)   How far did the train travel in the first half of the trip? Include units of measure.

(b)   What was the acceleration at $t = 3$ hours? Include units of measure.

(c)   If the train has returned to its starting point at $t = 10$ hours, estimate the value of A to the nearest whole number. Show how you arrived at your answer.

Section II Part A: Graphing Calculator MAY BE USED.

2. The temperature $T(x)$, in °F, in a small office building without air conditioning is given by $T(x) = 73 - 14\cos\left(\dfrac{\pi(x - 3.4)}{12}\right)$, where x is the time elapsed since midnight, $0 \le x \le 24$.

To cool the building, the air conditioning is turned on when the temperature first reaches the desired temperature T_0 and left on until the office closes at 6:00 p.m. (that is, when $x = 18$).

The cost per day, in dollars, of cooling is given by $C(x) = 0.16 \displaystyle\int_x^{18} (T(x) - T_0)\ dx$, for $T(x) \ge T_0$.

(a) Estimated to the nearest half-hour, at what time will the temperature first reach 70°F?

(b) Estimated to the nearest half-hour, at what time will the temperature first reach 77°F?

(c) What is the cost per day of cooling the office if the desired temperature is 70°F? Show your reasoning.

(d) How much is saved per day if the desired temperature is raised to 77°F?

(a) Estimated to the nearest half-hour, at what time will the temperature first reach 70°F?

(b)   Estimated to the nearest half-hour, at what time will the temperature first reach 77°F?

(c)   What is the cost per day of cooling the office if the desired temperature is 70°F? Show your reasoning.

(d)   How much is saved per day if the desired temperature is raised to 77°F?

Section II Part A: Graphing Calculator MAY BE USED.

3. This problem deals with functions defined by $f(x) = x^3 - 3bx$ with $b > 0$.

    (a) In the viewing window provided below, graph the members of the family $f(x) = x^3 - 3bx$ with $b = 1$, $b = 2$, and $b = 3$. Label each graph.

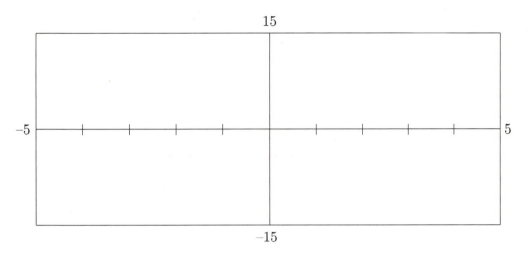

    (b) Find the x- and y-coordinates of the relative maximum points of f in terms of b.

    (c) Find the x- and y-coordinates of the relative minimum points of f in terms of b.

    (d) Show that for all values of $b > 0$, the relative maximum and minimum points lie on a function of the form $y = -ax^3$ and find the numerical value of a.

    (a) In the viewing window provided below, graph the members of the family $f(x) = x^3 - 3bx$ with $b = 1$, $b = 2$, and $b = 3$. Label each graph.

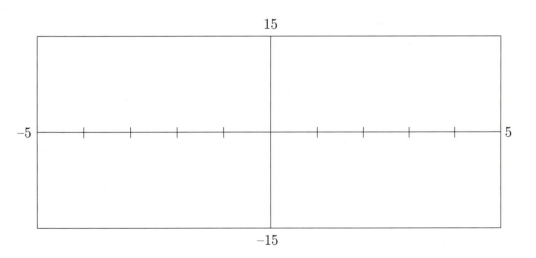

(b)   Find the x- and y-coordinates of the relative maximum points of f in terms of b.

(c)   Find the x- and y-coordinates of the relative minimum points of f in terms of b.

(d)   Show that for all values of $b > 0$, the relative maximum and minimum points lie on a function of the form $y = -ax^3$ and find the numerical value of a.

Section II Part B: Graphing Calculator MAY **NOT** BE USED.

4. Consider the differential equation $\dfrac{dy}{dx} = \dfrac{y - y^2}{x}$ for all $x \neq 0$.

(a) Verify that $y = \dfrac{x}{x + C}$, $x \neq -C$, is a general solution for the given differential equation.

(b) Show that all solutions with $C > 0$ contain $(0, 0)$.

(c) What is the value of C for which the particular solution will include the point $(-1, -1)$ and find the value of $\dfrac{dy}{dx}$ at $(0, 0)$ for this solution.

(d) The slope field for the given differential equation is provided. Sketch the particular solution that contains the point $(-1, -1)$.

(a) Verify that $y = \dfrac{x}{x + C}$, $x \neq -C$, is a general solution for the given differential equation.

(b)   Show that all solutions with $C > 0$ contain $(0, 0)$.

(c)   What is the value of C for which the particular solution will include the point $(-1, -1)$ and find the value of $\dfrac{dy}{dx}$ at $(0, 0)$ for this solution.

(d)   The slope field for the given differential equation is provided. Sketch the particular solution that contains the point $(-1, -1)$.

Section II Part B: Graphing Calculator MAY **NOT** BE USED.

5. A particle moves along the curve defined by the parametric equations $x(t) = 2t$ and $y(t) = 36 - t^2$ for time t, $0 \leq t \leq 6$. A laser light on the particle points in the direction of motion and shines on the x-axis.

(a) What is the velocity vector of the particle?

(b) Write an equation of the line tangent to the graph at $(2t, 36 - t^2)$ in terms of t and x.

(c) Express the x-coordinate of the point on the x-axis that the light hits as a function of t.

(d) At what time t is the light moving along the x-axis with the slowest speed? Justify your answer.

(a) What is the velocity vector of the particle?

(b) Write an equation of the line tangent to the graph at $(2t, 36 - t^2)$ in terms of t and x.

(c)   Express the x-coordinate of the point on the x-axis that the light hits as a function of t.

(d)   At what time t is the light moving along the x-axis with the slowest speed? Justify your answer.

6.

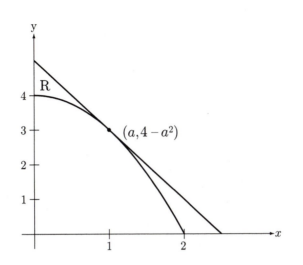

Let $f(x) = 4 - x^2$. Let R be the region in the first quadrant bounded by the graphs of f, its tangent line at $x = a$, and the y axis.

(a)   Write an equation of the tangent line in terms of a.

(b)   Write an integral expression that gives the area of the region R.

(c)   Write an expression for the area of the region R in terms of a, without an integral sign.

(d)   Find the average value of the area of the regions in terms of a.

(a)   Write an equation of the tangent line in terms of a.

(b)   Write an integral expression that gives the area of the region R.

(c)   Write an expression for the area of the region R in terms of a, without an integral sign.

(d)   Find the average value of the area of the regions in terms of a.

Sample Examination III

Directions: Solve each of the following problems, using the available space for scratchwork. After examining the form of the choices, decide which is the best of the choices given. Do not spend too much time on any one problem. Calculators may NOT be used on this part of the exam.

In this test: Unless otherwise specified, the domain of a function f is assumed to be the set of all real numbers x for which $f(x)$ is a real number.

1. The area of the region between the graph of $y = 3x^2 + 2x$ and the x-axis from $x = 1$ to $x = 3$ is

   (A) 36

   (B) 34

   (C) 31

   (D) 26

   (E) 12

   Answer

2. The graph in the xy-plane represented by $x = \cos t$ and $y = 1 - \cos 2t$ for $-\infty < t < \infty$ is

   (A) part of a hyperbola

   (B) part of a parabola

   (C) an ellipse

   (D) a semicircle

   (E) a straight line

   Answer

71

3. If $y = 2xe^{-x}$, then the graph of y has a point of inflection at $x =$

(A) −2

(B) 0

(C) 1

(D) 2

(E) 4

Answer

4. A particle moves in the xy-plane so that at any time t, $t > 0$, its coordinates are $x = e^t \sin t$ and $y = e^t \cos t$. At $t = \pi$, its velocity vector is

(A) $(e^\pi, -e^\pi)$

(B) $(0, -e^\pi)$

(C) $(-e^\pi, e^\pi)$

(D) (e^π, e^π)

(E) $(-e^\pi, -e^\pi)$

Answer

5. A particle moves along a straight line so that its velocity is given by $v(t) = t^2$. How far does the particle travel between $t = 1$ and $t = 3$?

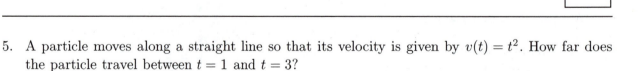

(A) $\frac{1}{3}$ (B) 8   (C) $\frac{26}{3}$ (D) 26   (E) 27

Answer

6. The derivative of $(4x^2)\cos(x)$ is

(A) $-8x\sin(x)$

(B) $8x\cos(x) - 4x^2\sin(x)$

(C) $8x\cos(x) + 4x^2\sin(x)$

(D) $-8x\cos(x) - 4x^2\sin(x)$

(E) $-8x\cos(x) + 4x^2\sin(x)$

Answer

7.

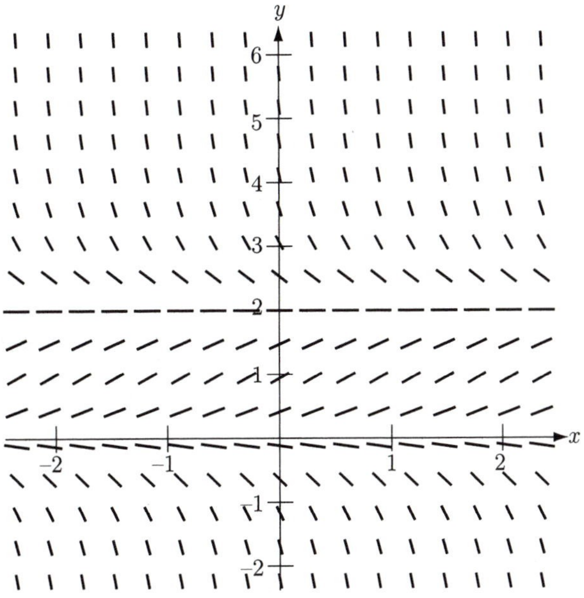

The slope field for a differential equation $\dfrac{dy}{dx} = f(y)$ is shown in the figure above.

Which statement is true about $y(x)$?

I. If $y(0) > 2$, then $\lim\limits_{x\to\infty} y(x) \approx 2$

II. If $0 < y(0) < 2$, then $\lim\limits_{x\to\infty} y(x) \approx 2$

III. If $y(0) < 0$, then $\lim\limits_{x\to\infty} y(x) \approx 2$

(A) I only   (B) II only   (C) III only   (D) I and II only   (E) I, II, and III

Answer

8. $\displaystyle\int_1^3 \frac{x}{x^2+1}\,dx =$

   (A) $\ln 5$

   (B) $\ln 10$

   (C) $2\ln 2$

   (D) $\frac{1}{2}\ln 5$

   (E) $\ln\left(\frac{5}{2}\right)$

Answer

9. $\displaystyle\frac{d}{dx}\ln\left(\frac{1}{x^2-1}\right) =$

   (A) x^2-1

   (B) $-\dfrac{2x}{x^2-1}$

   (C) $\dfrac{2x}{x^2-1}$

   (D) $2x^3-2x$

   (E) $2x-2x^3$

Answer

10. $\int_1^\infty x^{-\frac{5}{4}}\, dx$ is

   (A)   4

   (B) $\frac{5}{4}$

   (C) $\frac{1}{4}$

   (D) -4

   (E)   nonexistent

Answer

11. If the substitution $u = x^2 + 1$ is made, the integral $\int_0^2 \frac{x^3}{x^2 + 1}\, dx =$

   (A) $\int_1^5 \frac{u-1}{2u}\, du$

   (B) $\int_0^2 \frac{u-1}{2u}\, du$

   (C) $\int_1^5 \frac{(u-1)^{\frac{3}{2}}}{2u}\, du$

   (D) $\int_1^5 \frac{2(u-1)}{u}\, du$

   (E) $\int_0^2 \frac{2(u-1)}{u}\, du$

Answer

12. An equation of the line tangent to the curve $y = \dfrac{kx + 8}{k + x}$ at $x = -2$ is $y = x + 4$. What is the value of k?

   (A) -3 (B) -1 (C) 1   (D) 3   (E) 4

   Answer

13. For what positive value of k does $\displaystyle\int_0^k (4kx - 5k)\, dx = k^2$?

   (A) 1   (B) 2   (C) 3   (D) 4   (E) 5

   Answer

14. $\displaystyle\sum_{k=0}^{\infty} \left(-\dfrac{\pi}{3}\right)^k$ is

   (A) $\dfrac{1}{1 - \frac{\pi}{3}}$

   (B) $\dfrac{\frac{\pi}{3}}{1 - \frac{\pi}{3}}$

   (C) $\dfrac{1}{1 + \frac{\pi}{3}}$

   (D) $\dfrac{\frac{\pi}{3}}{1 + \frac{\pi}{3}}$

   (E) divergent

   Answer

15. If $y = 5^{(x^3-2)}$, then $\dfrac{dy}{dx} =$

(A) $(x^3 - 2)5^{(x^2-3)}$

(B) $3x^2(\ln 5)5^{(x^3-2)}$

(C) $(3x^2)5^{(x^3-2)}$

(D) $(\ln 5)5^{(x^3-2)}$

(E) $x^3(\ln 5)5^{(x^3-2)}$

Answer

16. If n is a positive integer, then $\displaystyle\lim_{n\to+\infty} \frac{1}{n}\left(\sin\frac{\pi}{n} + \sin\frac{2\pi}{n} + \cdots + \sin\frac{n\pi}{n} \right)$ is

(A) 0

(B) $\dfrac{2}{\pi}$

(C) $\dfrac{\pi}{2}$

(D) 2

(E) 2π

Answer

17.

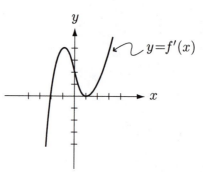

The graph of the <u>derivative</u> of f is shown in the figure above. Which of the following could be the graph of f?

(A)

(B)

(C)

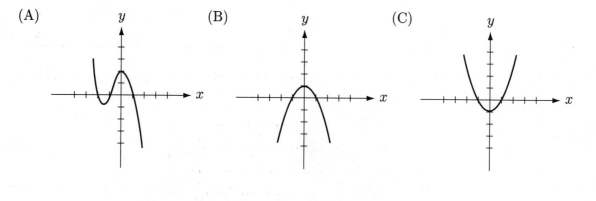

(D)

(E)

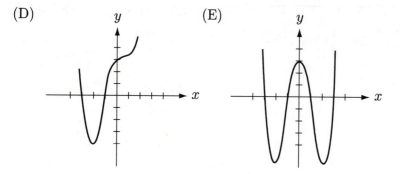

Answer

18. If $x = t - t^2$ and $y = \sqrt{2t + 5}$, then $\dfrac{dy}{dx}$ at $t = 2$ is

(A) $-\dfrac{1}{9}$ (B) -1 (C) $\dfrac{3}{2}$ (D) -9 (E) $-\dfrac{1}{18}$

Answer

19. Let $f(x)$ be a differentiable function defined only on the interval $-2 \le x \le 10$. The table below gives the value of $f(x)$ and its derivative $f'(x)$ at several points of the domain.

x	-2	0	2	4	6	8	10
$f(x)$	26	27	26	23	18	11	2
$f'(x)$	1	0	-1	-2	-3	-4	-5

The line tangent to the graph of $f(x)$ and parallel to the segment between the endpoints intersects the y-axis at the point

(A) $(0, 27)$

(B) $(0, 28)$

(C) $(0, 31)$

(D) $(0, 36)$

(E) $(0, 43)$

Answer

20. At each point (x, y) on a certain curve, the slope of the curve is $4xy$. If the curve contains the point $(0, 4)$, then its equation is

(A) $y = e^{2x^2} + 4$

(B) $y = e^{2x^2} + 3$

(C) $y = 2x^2 + 4$

(D) $y^2 = 4x^2 + 16$

(E) $y = 4e^{2x^2}$

Answer

21. $\displaystyle\int \frac{dx}{x^2 - 9} =$

(A) $\dfrac{1}{3}\ln\left|\dfrac{x+3}{x-3}\right| + C$

(B) $\ln|x^2 - 9| + C$

(C) $\dfrac{1}{6}\ln|(x-3)(x+3)| + C$

(D) $\dfrac{1}{6}\ln\left|\dfrac{x+3}{x-3}\right| + C$

(E) $\dfrac{1}{6}\ln\left|\dfrac{x-3}{x+3}\right| + C$

Answer

22. Which of the following definite integrals gives the length of the graph of $y = e^{(e^x)}$ between $x = 0$ and $x = 1$?

(A) $\displaystyle\int_0^1 \sqrt{1 + e^{2(x+e^x)}}\, dx$

(B) $\displaystyle\int_0^1 \sqrt{1 + e^{(x+e^x)}}\, dx$

(C) $\displaystyle\int_0^1 \sqrt{1 + e^{4x}}\, dx$

(D) $\displaystyle\int_0^1 \sqrt{1 + e^{2e^x}}\, dx$

(E) $\displaystyle\int_0^1 \sqrt{e^{(e^x)} + e^{(x+e^x)}}\, dx$

Answer

23. If the rate of change of a number x with respect to time, t, is x, what is the rate of change of the reciprocal of the number when $x = \frac{1}{4}$?

(A) -16

(B) -4

(C) $-\frac{1}{48}$

(D) $\frac{1}{48}$

(E) 4

Answer

24. If $f(x) = \sqrt{e^{2x} + 1}$, then $f'(0) =$

(A) $-\dfrac{\sqrt{2}}{2}$ (B) $\dfrac{\sqrt{2}}{4}$ (C) $\dfrac{\sqrt{2}}{2}$ (D) 1 (E) $\sqrt{2}$

Answer

25. If $x^2 y + yx^2 = 6$, then $\dfrac{d^2 y}{dx^2}$ at the point $(1, 3)$ is

(A) -18

(B) -6

(C) 6

(D) 12

(E) 18

Answer

26. The base of a solid is the region in the first quadrant bounded by the line $x + 2y = 4$ and the coordinate axes. What is the volume of the solid if every cross section perpendicular to the x-axis is a semicircle?

(A) $\dfrac{2\pi}{3}$ (B) $\dfrac{4\pi}{3}$ (C) $\dfrac{8\pi}{3}$ (D) $\dfrac{32\pi}{3}$ (E) $\dfrac{64\pi}{3}$

Answer

27. Which of the following series are convergent?

$$\text{I.} \quad \sum_{n=1}^{\infty}(-1)^{n+1}$$

$$\text{II.} \quad \sum_{n=1}^{\infty}(-1)^{n+1}(n)$$

$$\text{III.} \quad \sum_{n=1}^{\infty}\left(\frac{1+n}{n}\right)^{n}$$

(A) None

(B) II only

(C) III only

(D) I and II only

(E) I and III only

Answer

28. If $a > 0$ and $\lim_{x \to a} \dfrac{x^4 - a^4}{x^2 - a^2} = 16$, then $a =$

(A) 2

(B) $2\sqrt{2}$

(C) 4

(D) $4\sqrt{2}$

(E) 8

Answer

<u>Section I Part B</u>

<u>Directions</u>: Solve each of the following problems, using the available space for scratchwork. After examining the form of the choices, decide which is the best of the choices given. Do not spend too much time on any one problem. A graphing calculator is required for some questions on this part of the examination.

<u>In this test</u>:

(1)   The exact numerical value of the correct answer does not always appear among the choices given. When this happens, select from among the choices the number that best approximates the exact numerical value.

(2)   Unless otherwise specified, the domain of a function f is assumed to be the set of all real numbers x for which $f(x)$ is a real number.

29. A particle moves in a plane so that its position at any time, θ, $0 \leq \theta \leq 8$, is given by the polar equation $r(\theta) = 5(1 + \cos\theta)$. When does the particle's distance from the origin change from decreasing to increasing?

(A) $\theta = 0$ only

(B) $\theta = \pi$ only

(C) $\theta = 2\pi$ only

(D) $\theta = 0$ and π

(E) $\theta = \pi$ and 2π

Answer

30. The power series $\cos(x) = 1 - \dfrac{x^2}{2!} + \dfrac{x^4}{4!} - \dfrac{x^6}{6!} \cdots$ converges for all real numbers. For values in the interval $[0, \frac{\pi}{2}]$, what is the minimum number of terms of the power series necessary to approximate the value of $\cos(x)$ with an error whose absolute value is less than 0.0001?

(A) 4   (B) 5   (C) 6   (D) 7   (E) 8

Answer

31. If n is a positive integer, how many times does the function $f(x) = x^2 + 5\cos x$ change concavity in the interval $0 \le x \le 2\pi n$?

(A) 0

(B) 1

(C) 2

(D) n

(E) $2n$

Answer

32. A function $f(x)$ has a vertical asymptote at $x = 2$. The derivative of $f(x)$ is positive for all $x \ne 2$. Which of the following statements are true?

     I. $\displaystyle\lim_{x \to 2} f(x) = +\infty$

     II. $\displaystyle\lim_{x \to 2^-} f(x) = +\infty$

     III. $\displaystyle\lim_{x \to 2^+} f(x) = +\infty$

(A) I only

(B) II only

(C) III only

(D) I and II only

(E) I, II, and III

Answer

33. If the derivative of a function f is given by $f'(x) = \sin(x^x)$, then how many critical points does the function $f(x)$ have on the interval $[0.2, 2.6]$?

    (A) None

    (B) One

    (C) Two

    (D) Three

    (E) Four

Answer

34. Which of the following functions grows faster than e^x as $x \to \infty$?

    (A) x^4

    (B) $\ln x$

    (C) e^{-x}

    (D) 3^x

    (E) $\frac{1}{2}e^x$

Answer

35. The second derivative of a function is given by $f''(x) = 0.5 + \cos x - e^{-x}$. How many points of inflection does the function $f(x)$ have on the interval $0 \le x \le 20$?

   (A) None

   (B) Three

   (C) Six

   (D) Seven

   (E) Ten

Answer

36. The area of the region enclosed by the polar curve $r = 2(\cos(\theta) + \sin(\theta))$ is

   (A) 1

   (B) 2

   (C) π

   (D) 2π

   (E) 4π

Answer

37. Let f be a function that is everywhere differentiable. The table below provides information about $f(x)$ and its first, second, and third derivatives for selected values of x.

x	$f(x)$	$f'(x)$	$f''(x)$	$f'''(x)$
0	4	2	1	0.50
1	5.15	2.50	1.25	0.75
2	7.20	3.50	1.75	0.85
3	10.50	5.25	2.10	1.00

Which of the following best approximates $f(2.2)$?

(A) $4 + 0.2 + 3.50(0.2)^2 + 0.50(0.2)^3$

(B) $4 + 2.2 + \frac{1}{2}(2.2)^2 + \frac{0.50}{6}(2.2)^3$

(C) $7.20 + 3.50(2.2) + 1.75(2.2)^2 + 0.85(2.2)^3$

(D) $7.20 + 3.50(2.2) + \frac{1.75}{2}(2.2)^2 + \frac{0.85}{6}(2.2)^3$

(E) $7.20 + 3.50(0.2) + \frac{1.75}{2}(0.2)^2 + \frac{0.85}{6}(0.2)^3$

Answer

38. Let E be the error when the Taylor polynomial $T(x) = x - \dfrac{x^3}{3!}$ is used to approximate $f(x) = \sin x$ at $x = 0.5$. Which of the following is true?

(A) $|E| < 0.0001$

(B) $0.0001 < |E| < 0.0003$

(C) $0.0003 < |E| < 0.0005$

(D) $0.0005 < |E| < 0.0007$

(E) $0.0007 < |E|$

Answer

39. How many zeros does the function $y = \sin(\ln x)$ have for $0 < x \leq 1$?

    (A) One

    (B) Two

    (C) Three

    (D) Four

    (E) More than four

    Answer

---

40. Let $f(x) = x^3 - 7x^2 + 25x - 39$ and let g be the inverse function of f. What is the value of $g'(0)$?

    (A) $-\dfrac{1}{25}$ (B) $\dfrac{1}{25}$ (C) $\dfrac{1}{10}$ (D) 10   (E) 25

    Answer

---

41. If $\dfrac{dy}{dx} = xy - y^2$ and $y(1) = 3$, then $y(2) \approx$

    (A) -3

    (B) -1

    (C) 0

    (D) 3

    (E) 9

    Answer

42. A rectangle is to be inscribed in a semicircle of radius 8, with one side lying on the diameter of the circle. What is the maximum possible area of the rectangle?

   (A) $4\sqrt{2}$

   (B) $8\sqrt{2}$

   (C) 32

   (D) $32\sqrt{2}$

   (E) 64

Answer

43. $\int x^n \sin x \, dx =$

   (A) $-x^n \cos x + n \int x^{n-1} \cos x \, dx$

   (B) $-x^n \cos x - n \int x^{n-1} \cos x \, dx$

   (C) $x^n \cos x - n \int x^{n-1} \cos x \, dx$

   (D) $x^n \cos x + n \int x^{n-1} \cos x \, dx$

   (E) $-x^n \cos x + n \int x^{n-1} \sin x \, dx$

Answer

44. If $\dfrac{dy}{dt} = \dfrac{2y}{t(t+2)}$ for $t > 0$ and $y = 1$ when $t = 1$, then when $t = 2$, $y =$

(A) 0

(B) $\dfrac{1}{2}$

(C) $\dfrac{2}{3}$

(D) 1

(E) $\dfrac{3}{2}$

Answer

45. The line $y = mx + b$ with $b \geq 2$ is tangent to the graph of $f(x) = -2(x-2)^2 + 2$ at a point in the first quadrant. What are all possible values of b?

(A) $b = 2$ only

(B) $2 \leq b < 10$

(C) $2 \leq b < 12$

(D) $2 \leq b < 14$

(E) $2 \leq b < 20$

Answer

SECTION II - FREE-RESPONSE QUESTIONS
GENERAL INSTRUCTIONS

You may wish to look over the problems before starting to work on them, since it is not expected that everyone will be able to complete all parts of all problems. All problems are given equal weight, but the parts of a particular problem are not necessarily given equal weight.

- YOU SHOULD WRITE ALL WORK FOR EACH PART OF EACH PROBLEM IN THE SPACE PROVIDED FOR THAT PART. Be sure to write clearly and legibly. If you make an error, you may save time by crossing it out rather than trying to erase it. Erased or crossed-out work will not be graded.

- Show all your work. Clearly label any functions, graphs, tables, or other objects that you use. You will be graded on the correctness and completeness of your methods as well as your answers. Answers without supporting work may not receive credit.

- Justifications require that you give mathematical (noncalculator) reasons.

- Your work must be expressed in standard mathematical notation rather than calculator syntax. For example, $\int_1^5 x^2 \, dx$ may not be written as fnInt(X^2, X, 1, 5).

- Unless otherwise specified, answers (numeric or algebraic) need not be simplified.

- If you use decimal approximations in calculations, you will be graded on accuracy. Unless otherwise specified, your final answers should be accurate to three places after the decimal point.

- Unless otherwise specified, the domain of a function f is assumed to be the set of all real numbers x for which $f(x)$ is a real number.

SECTION II PART A: 45 Minutes, Questions 1,2,3

During the timed portion for Part A, you may work only on the problems in Part A. Write your solution to each part of each problem in the space provided for that part.

On Part A, you are permitted to use your calculator to solve an equation, find the derivative of a function at a point, or calculate the value of a definite integral. However, you must clearly indicate the setup of your problem, namely the equation, function, or integral you are using. If you use other built-in features or programs, you must show the mathematical steps necessary to produce your results.

Do not go on to Part B until you are told to do so.

SECTION II PART B: 45 Minutes, Questions 4,5,6

Write your solution to each part of each problem in the space provided for that part. During the timed portion for Part B, you may continue to work on the problems in Part A without the use of any calculator.

1. A particle moves along the x-axis so that at any time t, $0.1 < t < 0.3$, its velocity is given by
$v(t) = \sin\left(\dfrac{1}{t}\right)$.

  (a) Find an expression for the acceleration of the particle at any time t in the given interval.

  (b) When is the velocity the greatest? Justify your answer.

  (c) Sketch the graph of the velocity as a function of time t in the viewing window provided
below.

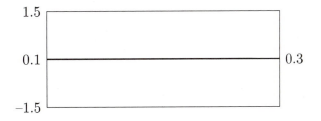

Viewing Window
$[0.1, 0.3]$ by $[-1.5, 1.5]$

  (d) During the given interval, does the particle spend more time moving to the left or to the
right? Justify your answer.

_____

  (a) Find an expression for the acceleration of the particle at any time t in the given interval.

(b)   When is the velocity the greatest? Justify your answer.

(c)   Sketch the graph of the velocity as a function of time t in the viewing window provided below.

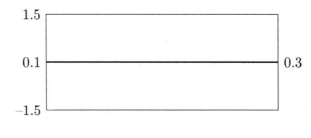

Viewing Window
$[0.1, 0.3]$ by $[-1.5, 1.5]$

(d)   During the given interval, does the particle spend more time moving to the left or to the right? Justify your answer.

> Section II Part A: Graphing Calculator MAY BE USED.

2. Sand is being poured into a bin that is initially empty. During the workday, for $0 \leq t \leq 9$ hours, the sand pours into the bin at the rate given by

$$S(t) = \frac{5000}{t^3 + 50} \text{ cubic meters per hour.}$$

After one hour, for $1 \leq t \leq 9$, sand is removed from the bin at the rate of

$$R(t) = 23.967\sqrt{t} \text{ cubic meters per hour.}$$

(a) How much sand is put into the bin during the workday? Include units of measure.

(b) Find $S(6) - R(6)$; include units of measure. Explain what this number means in the context of the problem.

(c) Explain why the maximum amount of sand in the bin occurs when $S(t) = R(t)$.

(d) How much sand is in the bin at the end of the workday? Indicate units of measure.

(a) How much sand is put into the bin during the workday? Include units of measure.

(b) Find $S(6) - R(6)$; include units of measure. Explain what this number means in the context of the problem.

(c)   Explain why the maximum amount of sand in the bin occurs when $S(t) = R(t)$.

(d)   How much sand is in the bin at the end of the workday? Indicate units of measure.

Section II Part A: Graphing Calculator MAY BE USED.

3. It is estimated that at the current rate of consumption, r gallons per year, the oil supply of the earth will last 200 years. However, the rate of consumption, $R(t)$, is increasing at the rate of 5% per year; that is $\dfrac{dR}{dt} = 0.05R$.

   (a) In terms of r, how many gallons of oil are currently available?

   (b) Use the given differential equation to find $R(t)$.

   (c) If no additional oil is discovered, how long, to the nearest year, will the current oil supply actually last? Show how you arrived at your solution.

   (a) In terms of r, how many gallons of oil are currently available?

(b)   Use the given differential equation to find $R(t)$.

(c)   If no additional oil is discovered, how long, to the nearest year, will the current oil supply actually last? Show how you arrived at your solution.

4.

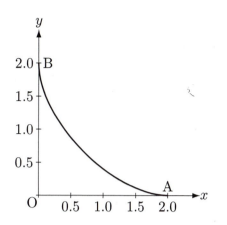

Let the coordinate axes represent two highways that meet at right angles. In order to safely connect the two roads, a curved road is to be built from point A to point B as shown in the figure above. The parametric equations of the connecting road are

$$x(t) = 2\cos^3 t \text{ and } y(t) = 2\sin^3 t \text{ for } 0 \le t \le \frac{\pi}{2}, \text{ where } x(t) \text{ and } y(t) \text{ are in miles.}$$

In order to make the turns safe, the slope of the curved road should be the same as the slope of each highway where they meet.

(a) Find $\dfrac{dy}{dx}$ in terms of t.

(b) Show that the slope of the curved road and the slope of each highway are the same at points A and B.

(c) Set up but do not evaluate an integral expression which gives the length of the curved road from A to B.

(a) Find $\dfrac{dy}{dx}$ in terms of t.

(b)   Show that the slope of the curved road and the slope of each highway are the same at points A and B.

(c)   Set up but do not evaluate an integral expression which gives the length of the curved road from A to B.

Section II Part B: Graphing Calculator MAY **NOT** BE USED.

5. Let $f(x)$ be the function defined by the power series

$$1 - \frac{1}{3}(x-2) + \frac{1}{9}(x-2)^2 - \frac{1}{27}(x-2)^3 + \ldots + (-1)^n \frac{(x-2)^n}{3^n} + \ldots$$

(a) For what values of x does the series converge?

(b) Write the first four terms and the general term of the power series for $f'(x)$ centered at $x = 2$ and find $f'(2)$.

(c) Write an equation of the tangent line of $f(x)$ at the point where $x = 2$.

(d) Near $x = 2$, does the tangent line lie above the curve or below the curve? Justify your answer.

(a) For what values of x does the series converge?

(b) Write the first four terms and the general term of the power series for $f'(x)$ centered at $x = 2$ and find $f'(2)$.

(c)   Write an equation of the tangent line of $f(x)$ at the point where $x = 2$.

(d)   Near $x = 2$, does the tangent line lie above the curve or below the curve? Justify your answer.

6.

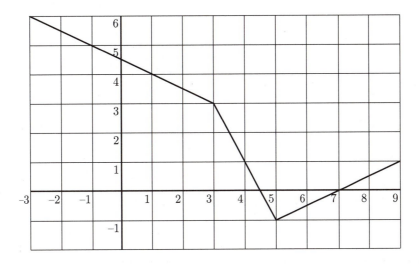

The graph of f

Let f be a function defined on the closed interval $[-3, 9]$. The graph of f, consisting of three line segments, is shown above. Let $g(x) = \int_0^x f(t)\, dt$.

(a) Find $g(4.5)$, $g'(4.5)$, and $g''(4.5)$.

(b) Find the average value of f on the interval $[-3, 5]$. Show the work that leads to your answer.

(c) Find the x-coordinate of any points of inflection of g. Justify your answer.

(d) Find the coordinates of all maximum points of g.

(a) Find $g(4.5)$, $g'(4.5)$, and $g''(4.5)$.

(b) Find the average value of f on the interval $[-3, 5]$. Show the work that leads to your answer.

(c) Find the x-coordinate of any points of inflection of g. Justify your answer.

(d) Find the coordinates of all maximum points of g.

Sample Examination IV

Directions: Solve each of the following problems, using the available space for scratchwork. After examining the form of the choices, decide which is the best of the choices given. Do not spend too much time on any one problem. Calculators may NOT be used on this part of the exam.

In this test: Unless otherwise specified, the domain of a function f is assumed to be the set of all real numbers x for which $f(x)$ is a real number.

1. If f is continuous for all real numbers, $\dfrac{dy}{dx} = f(x)$ and $y(2) = 4$, then $y(x) =$

   (A) $4 + \displaystyle\int_2^x f'(t)\, dt$

   (B) $4 + \displaystyle\int_2^x f(t)\, dt$

   (C) $\displaystyle\int_2^x f(t)\, dt - 4$

   (D) $f(x) - f(2)$

   (E) None of the above.

   Answer

2. For what value(s) of x does $4x^6 - 8x^3 + 18$ have a relative minimum?

   (A) -1 only

   (B) 0 only

   (C) 1 only

   (D) 0 and 1 only

   (E) $-1, 0$ and 1

   Answer

3. A curve is given parametrically by the equations $x = 3 - 4\sin t$ and $y = 4 + 3\cos t$ for $0 \le t \le 2\pi$. What are all points (x, y) at which the curve has a vertical tangent?

(A) $(-1, 4)$

(B) $(3, 7)$

(C) $(-1, 4)$ and $(7, 4)$

(D) $(3, 7)$ and $(3, 1)$

(E) $(4, -1)$ and $(4, 7)$

Answer

4. If $x = 2t^2$ and $y = t^3$, then $\dfrac{d^2 y}{dx^2}$ at $t = 3$ is

(A) $\dfrac{1}{16}$

(B) $\dfrac{9}{2}$

(C) $\dfrac{3}{4}$

(D) $\dfrac{1}{4}$

(E) $\dfrac{9}{4}$

Answer

5. What are all values of x for which $\displaystyle\sum_{n=1}^{\infty} \frac{2^n x^n}{n}$ converges?

(A) $|x| = \frac{1}{2}$ only

(B) $-\frac{1}{2} \le x \le \frac{1}{2}$

(C) $-\frac{1}{2} < x < \frac{1}{2}$

(D) $-\frac{1}{2} < x \le \frac{1}{2}$

(E) $-\frac{1}{2} \le x < \frac{1}{2}$

Answer

6. The position of a particle on the x-axis at time t, $t > 0$, is $\ln t$. The average velocity of the particle for $1 \le t \le e$ is

(A) 1

(B) $\dfrac{1}{e} - 1$

(C) $\dfrac{1}{e-1}$

(D) e

(E) $e - 1$

Answer

7. If the function $r = f(\theta)$ is continuous and nonnegative for $0 \le \alpha \le \theta \le \beta \le 2\pi$, then the area enclosed by the polar curve $r = f(\theta)$ and the lines $\theta = \alpha$ and $\theta = \beta$ is given by

(A) $\dfrac{1}{2} \displaystyle\int_{\alpha}^{\beta} f(\theta^2)\, d\theta$

(B) $\dfrac{1}{2} \displaystyle\int_{\alpha}^{\beta} f(\theta)\, d\theta$

(C) $\dfrac{1}{2} \displaystyle\int_{\alpha}^{\beta} \theta f(\theta^2)\, d\theta$

(D) $\dfrac{1}{2} \displaystyle\int_{\alpha}^{\beta} \theta f(\theta)\, d\theta$

(E) $\dfrac{1}{2} \displaystyle\int_{\alpha}^{\beta} [f(\theta)]^2\, d\theta$

Answer

8. An antiderivative of $2x\cos(2x)$ is

(A) $\sin(2x)$

(B) $x\sin(2x) + \cos(2x)$

(C) $x\cos(2x) + \frac{1}{2}\sin(2x)$

(D) $x\sin(2x) + \frac{1}{2}\cos(2x)$

(E) $\frac{1}{2}x\sin(2x) + \frac{1}{2}\cos(2x)$

Answer

9. $\displaystyle\int_{-3}^{3} |x + 2|\ dx =$

(A) 0

(B) 8

(C) 13

(D) 17

(E) 21

Answer

10. Let R be the region in the <u>fourth quadrant</u> enclosed by the x-axis and the curve $y = x^2 - 2kx$, where $k > 0$. If the area of the region R is 36, then the value of k is

(A) 2

(B) 3

(C) 4

(D) 6

(E) 9

Answer

11. $\displaystyle\lim_{h \to 0} \frac{1}{h} \int_{0}^{h} \frac{\sin^2 t}{t^2}\ dt =$

(A) 0   (B) $\dfrac{1}{2}$ (C) 1   (D) 2   (E) ∞

Answer

12.

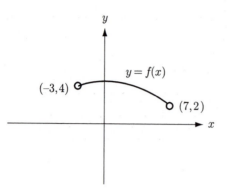

The graph of $y = f(x)$ on the closed interval $[-3, 7]$ is shown in the figure above. If f is continuous on $[-3, 7]$ and differentiable on $(-3, 7)$, then there exists a c, $-3 < c < 7$, such that

(A) $f(c) = 0$

(B) $f'(c) = -\dfrac{1}{5}$

(C) $f'(c) = \dfrac{1}{5}$

(D) $f'(c) = -5$

(E) $f'(c) = 5$

Answer

13. If $y = \dfrac{1-x}{x-1}$, then $\dfrac{dy}{dx} =$

(A) -1 (B) 0 (C) $\dfrac{-1}{x-1}$ (D) $\dfrac{-2}{x-1}$ (E) $\dfrac{-2x}{(x-1)^2}$

Answer

14. Let $f(x)$ be a differentiable function. The table below gives the value of $f(x)$ and $f'(x)$, the derivative of $f(x)$, at several values of x. If $g(x) = \frac{1}{f(x)}$, what is the value of $g'(2)$?

x	1	2	3	4
$f(x)$	–3	–8	–9	0
$f'(x)$	–5	–4	3	16

(A) $-\dfrac{1}{8}$

(B) 0

(C) $\dfrac{1}{16}$

(D) $\dfrac{1}{64}$

(E) 16

Answer

15. Which of the following series converge?

I. $\displaystyle\sum_{n=1}^{\infty} \left(1 - \frac{4}{3}\right)^n$ II. $\displaystyle\sum_{n=1}^{\infty} \left(1 + \frac{4}{3}\right)^n$ III. $\displaystyle\sum_{n=1}^{\infty} \left(1 + \frac{1}{n}\right)^n$

(A) I only

(B) II only

(C) III only

(D) I and II only

(E) I and III only

Answer

16. The Taylor Series of a function $f(x)$ about $x = 3$ is given by

$$f(x) = 1 + \frac{3(x-3)}{1!} + \frac{5(x-3)^2}{2!} + \frac{7(x-3)^3}{3!} + \cdots + \frac{(2n+1)(x-3)^n}{n!} + \cdots$$

What is the value of $f'''(3)$?

(A) 0

(B) $1.1\bar{6}$

(C) 2.5

(D) 5

(E) 7

Answer

17. If $f(x) = 15 - g(x)$ for $-2 \leq x \leq 2$, then $\int_{-2}^{2} (f(x) - g(x)) \ dx =$

(A) 60

(B) $2 \int_{-2}^{2} g(x) \ dx - 60$

(C) $2 \int_{-2}^{2} g(x) \ dx + 60$

(D) $60 - 4 \int_{0}^{2} g(x) \ dx$

(E) $60 - 2 \int_{-2}^{2} g(x) \ dx$

Answer

18. $\int_{2}^{4} \left(\dfrac{d}{dt}(3t^2 + 2t - 1) \right) \, dt =$

   (A) 12

   (B) 40

   (C) 46

   (D) 55

   (E) 66

   Answer

19. Let a_n, b_n, and c_n be sequences of positive numbers such that for all positive integers n, $a_n \le b_n \le c_n$. If $\displaystyle\sum_{n=1}^{\infty} b_n$ converges, then which of the following statements must be true?

   I. $\displaystyle\sum_{n=1}^{\infty} a_n$ converges

   II. $\displaystyle\sum_{n=1}^{\infty} c_n$ converges

   III. $\displaystyle\sum_{n=1}^{\infty} (a_n + b_n)$ converges

   (A) I only

   (B) II only

   (C) III only

   (D) I and III only

   (E) I, II, and III

   Answer

20. Let $f(x) = \begin{cases} 1 + e^{-x}, & 0 \le x \le 5 \\ 1 + e^{x-10}, & 5 < x \le 10 \end{cases}$

Which of the following statements are true?

    I. $f(x)$ is continuous for all values of x in the interval $[0, 10]$.

    II. $f'(x)$, the derivative of $f(x)$, is continuous for all values of x in the interval $[0, 10]$.

    III.   The graph of $f(x)$ is concave upwards for all values of x in the interval $[0, 10]$.

(A) I only

(B) II only

(C) III only

(D) I and III only

(E) I and II and III

Answer

21. A solid has a circular base of radius 3. If every plane cross section perpendicular to the x-axis is an equilateral triangle, then its volume is

(A) 36

(B) $12\sqrt{3}$

(C) $18\sqrt{3}$

(D) $24\sqrt{3}$

(E) $36\sqrt{3}$

Answer

22. If the substitution $u = 25 - x^2$ is made, the integral $\displaystyle\int_0^3 x\sqrt{25 - x^2}\,dx =$

(A) $\dfrac{1}{2}\displaystyle\int_0^3 \sqrt{u}\,du$

(B) $\dfrac{1}{2}\displaystyle\int_{25}^{16} \sqrt{u}\,du$

(C) $-\dfrac{1}{2}\displaystyle\int_0^3 \sqrt{u}\,du$

(D) $\dfrac{1}{2}\displaystyle\int_{16}^{25} \sqrt{u}\,du$

(E) $2\displaystyle\int_{16}^{25} \sqrt{u}\,du$

Answer

23. If $s_n = \left(\dfrac{(8-n)^{200}}{8^{n+2}}\right)\left(\dfrac{8^n}{(3-n^2)^{100}}\right)$, to what number does the sequence $\{s_n\}$ converge?

(A) $-\dfrac{1}{8}$

(B) $-\dfrac{1}{64}$

(C) 0

(D) $\dfrac{1}{64}$

(E) $\dfrac{1}{8}$

Answer

24. Which of the functions sketched below is increasing at a decreasing rate?

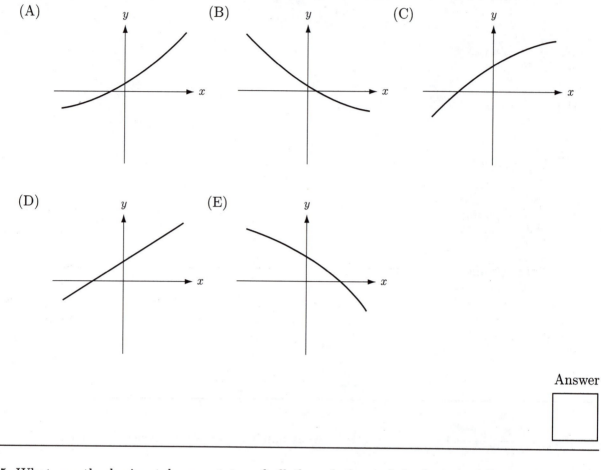

(A)   (B)   (C)

(D)   (E)

Answer

Answer

25. What are the horizontal asymptotes of all the solutions of the logistic differential equation $\frac{dy}{dx} = y\left(8 - \frac{y}{1,000}\right)$?

(A) $y = 0$ only

(B) $y = 8$ only

(C) $y = 8,000$ only

(D) $y = 0$ and $y = 8$

(E) $y = 0$ and $y = 8,000$

Answer

26. $\lim\limits_{h \to 0} \dfrac{\tan\left(2(x+h)\right) - \tan(2x)}{h}$ is

(A) 0

(B) $2\sec^2(2x)$

(C) $\sec^2(2x)$

(D) $2\cot(2x)$

(E) nonexistent

Answer

27. The slope field for the differential equation $\frac{dy}{dx} = f(x)$ is shown below for $-4 < x < 4$ and $-4 < y < 4$. Which of the following statements is true for all possible solutions of the differential equation?

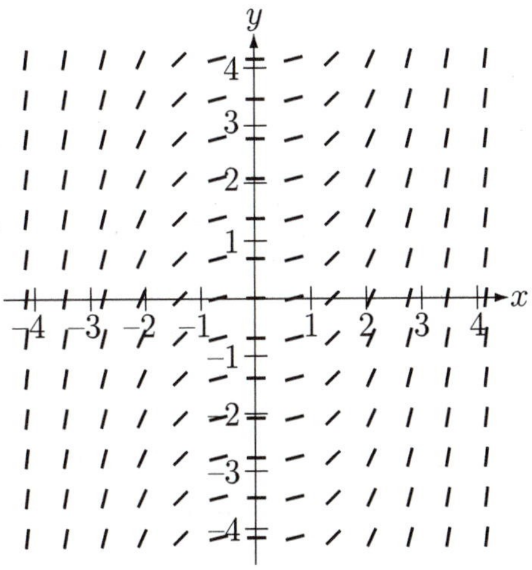

I.   For $x < 0$, all solution functions are decreasing.
II.   For $x > 0$, all solution functions are increasing.
III.   All solution functions level off near the y-axis.

(A) I only
(B) II only
(C) III only
(D) II and III only
(E) I, II, and III

Answer

28. At which of the three points on the graph of $y = f(x)$ in the figure below will $f'(x) < f''(x)$?

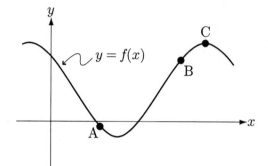

(A) A only

(B) B only

(C) C only

(D) A and B only

(E) A and C only

Answer

Section I Part B

Directions: Solve each of the following problems, using the available space for scratchwork. After examining the form of the choices, decide which is the best of the choices given. Do not spend too much time on any one problem. A graphing calculator is required for some questions on this part of the examination.

In this test:

(1) The exact numerical value of the correct answer does not always appear among the choices given. When this happens, select from among the choices the number that best approximates the exact numerical value.

(2) Unless otherwise specified, the domain of a function f is assumed to be the set of all real numbers x for which $f(x)$ is a real number.

29. If $e^{xy} = 2$, then at the point $(1, \ln 2)$, $\dfrac{dy}{dx} =$

(A) $-\ln 2$

(B) $2 \ln 2$

(C) $\ln 2$

(D) $-2e$

(E) $-4 \ln 2$

Answer

30. A particle moves along the x-axis so that its position at any time $t > 0$ is given by $x(t) = t^4 - 10t^3 + 29t^2 - 36t + 2$. For which value of t is the speed the greatest?

(A) $t = 1$

(B) $t = 2$

(C) $t = 3$

(D) $t = 4$

(E) $t = 5$

Answer

31.

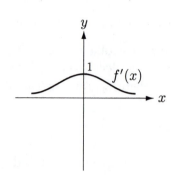

Suppose the derivative of f has the graph shown above.

Which of the following could be the graph of f?

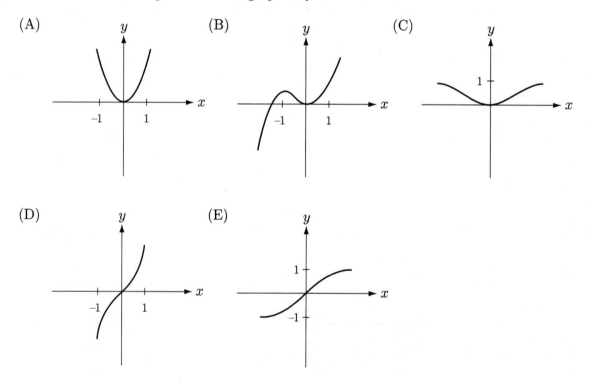

(A) (B) (C)

(D) (E)

Answer

32. The velocity vector of a particle moving in the xy-plane is given by $\vec{v} = (2\sin t, 3\cos t)$ for $t \geq 0$. At $t = 0$, the particle is at the point $(1, 1)$. What is the position vector at $t = 2$?

(A) $(3.832, 3.728)$

(B) $(1.832, -1.728)$

(C) $(1.819, -1.248)$

(D) $(1.735, -0.532)$

(E) $(0, 3)$

Answer

33. Using the subintervals $[1, 5]$, $[5, 8]$, and $[8, 10]$, what is the trapezoidal approximation to $\int_1^{10} (2 - \cos x)\, dx$?

(A) 17.126

(B) 17.129

(C) 17.155

(D) 18.147

(E) 19.385

Answer

34. The length of the curve $y = x^3$ from $(0,0)$ to $(1,1)$ is

   (A) 1.380

   (B) 1.414

   (C) 1.495

   (D) 1.548

   (E) 1.732

Answer

35. Let f be a function whose derivative is given by $f'(x) = \frac{x}{15} + \sin(e^{0.2x})$. How many relative maximum points does $f(x)$ have in the interval $0 < x < 12$?

   (A) None

   (B) One

   (C) Two

   (D) Three

   (E) More than three

Answer

36. Two cars start at the same place and at the same time. One car travels west at a constant velocity of 50 miles per hour and a second car travels south at a constant velocity of 60 miles per hour. Approximately how fast is the distance between them changing one-half hour later?

(A) 72 miles per hour

(B) 74 miles per hour

(C) 76 miles per hour

(D) 78 miles per hour

(E) 80 miles per hour

Answer

37.

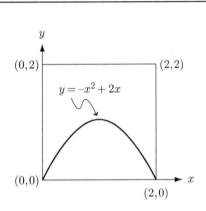

As shown in the figure above, a square with vertices $(0,0)$, $(2,0)$, $(2,2)$, and $(0,2)$ is divided into two regions by the graph of $y = -x^2 + 2x$. If a point is picked at random from inside the square, what is the probability that the point lies in the region <u>above</u> the parabola?

(A) $\frac{3}{4}$

(B) $\frac{5}{7}$

(C) $\frac{2}{3}$

(D) $\frac{5}{8}$

(E) $\frac{3}{5}$

Answer

38. If the derivative of a function f is given by $f'(x) = (x^2 + 5)^5$, how many points of inflection does the function f have?

(A) None

(B) One

(C) Two

(D) Three

(E) Four

Answer

39. Let $f(x)$ be a function that is continuous and differentiable for all x. The derivative of this function is given by the power series

$$f'(x) = 3x - \frac{9x^3}{2} + \frac{81x^5}{40} - \frac{3x^7}{60} + \cdots$$

If $f(0) = 2$, then $f(x) =$

(A) $0 + 3x - \frac{9x^3}{2} + \frac{81x^5}{40} - \frac{3x^7}{60} + \cdots$

(B) $2 + 3x - \frac{9x^3}{2} + \frac{81x^5}{40} - \frac{3x^7}{60} + \cdots$

(C) $\frac{3x^2}{2} - \frac{9x^4}{8} + \frac{27x^6}{80} - \frac{3x^8}{480} + \cdots$

(D) $2 - \frac{3x^2}{2} + \frac{9x^4}{8} - \frac{27x^6}{80} + \frac{3x^8}{480} + \cdots$

(E) $2 + \frac{3x^2}{2} - \frac{9x^4}{8} + \frac{27x^6}{80} - \frac{3x^8}{480} + \cdots$

Answer

40. Two particles, Alpha and Beta, race from the y-axis to the vertical line $x = 6\pi$. Alpha's position is given by the parametric equations $x_\alpha = (3t - 3\sin t)$, $y_\alpha = (3 - 3\cos t)$ and Beta's position is given by $x_\beta = (3t - 4\sin t)$, $y_\beta = (3 - 4\cos t)$ for $t \geq 0$. Which sentence best describes the race and its outcome?

(A) Alpha moves slower and loses.

(B) Alpha takes a shorter path and wins.

(C) Beta starts out in the wrong direction and loses.

(D) Beta moves faster but loses.

(E) Alpha and Beta tie.

Answer

41. If $\dfrac{dy}{dx} = e^x y$ and $y = 3$ when $x = 0$, then $y =$

(A) $3e^x$

(B) $3e^{e^x}$

(C) $\dfrac{1}{3}e^{e^x}$

(D) $\dfrac{3e^x}{e}$

(E) $\dfrac{3e^{e^x}}{e}$

Answer

42. For what values of x does the curve $y^2 - x^3 - 15x^2 = 10$ have horizontal tangent lines?

   (A) $x = -10$ only

   (B) $x = 0$ only

   (C) $x = 10$ only

   (D) $x = 0$ and $x = -10$ only

   (E) $x = -10$, $x = 0$, and $x = 10$

Answer

43. The amount of a radioactive substance decays according to the equation $\dfrac{dy}{dt} = ky$ where k is a constant and time, t, is measured in days. If half of the amount present will decay in 20 days, what is the value of k?

   (A) -13.066

   (B) -6.021

   (C) -0.693

   (D) -0.035

   (E) -0.015

Answer

44.

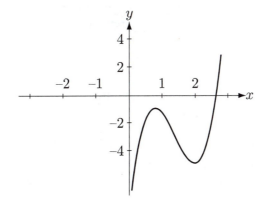

The graph above shows a function f with a relative minimum at $x = 2$. The approximation of $f(x)$ near $x = 2$ by the second-degree Taylor polynomial centered about $x = 2$ is given by $a + b(x - 2) + c(x - 2)^2$.

Which of the following is true about a, b, and c?

(A) $a < 0, b = 0, c > 0$

(B) $a > 0, b = 0, c < 0$

(C) $a < 0, b < 0, c < 0$

(D) $a < 0, b > 0, c > 0$

(E) $a > 0, b = 0, c > 0$

Answer

45. The solution of the differential equation $\frac{dy}{dx} = -\frac{x^2}{y}$ contains the point $(3, -2)$. Using Euler's method with $\Delta x = -0.3$, what is the approximate value of y when $x = 2.7$?

(A) -2.98

(B) -3.00

(C) -3.08

(D) -3.25

(E) -3.35

Answer

SECTION II - FREE-RESPONSE QUESTIONS
GENERAL INSTRUCTIONS

You may wish to look over the problems before starting to work on them, since it is not expected that everyone will be able to complete all parts of all problems. All problems are given equal weight, but the parts of a particular problem are not necessarily given equal weight.

- YOU SHOULD WRITE ALL WORK FOR EACH PART OF EACH PROBLEM IN THE SPACE PROVIDED FOR THAT PART. Be sure to write clearly and legibly. If you make an error, you may save time by crossing it out rather than trying to erase it. Erased or crossed-out work will not be graded.

- Show all your work. Clearly label any functions, graphs, tables, or other objects that you use. You will be graded on the correctness and completeness of your methods as well as your answers. Answers without supporting work may not receive credit.

- Justifications require that you give mathematical (noncalculator) reasons.

- Your work must be expressed in standard mathematical notation rather than calculator syntax. For example, $\int_{1}^{5} x^2 \, dx$ may not be written as fnInt(X^2, X, 1, 5).

- Unless otherwise specified, answers (numeric or algebraic) need not be simplified.

- If you use decimal approximations in calculations, you will be graded on accuracy. Unless otherwise specified, your final answers should be accurate to three places after the decimal point.

- Unless otherwise specified, the domain of a function f is assumed to be the set of all real numbers x for which $f(x)$ is a real number.

SECTION II PART A: 45 Minutes, Questions 1,2,3

During the timed portion for Part A, you may work only on the problems in Part A. Write your solution to each part of each problem in the space provided for that part.

On Part A, you are permitted to use your calculator to solve an equation, find the derivative of a function at a point, or calculate the value of a definite integral. However, you must clearly indicate the setup of your problem, namely the equation, function, or integral you are using. If you use other built-in features or programs, you must show the mathematical steps necessary to produce your results.

Do not go on to Part B until you are told to do so.

SECTION II PART B: 45 Minutes, Questions 4,5,6

Write your solution to each part of each problem in the space provided for that part. During the timed portion for Part B, you may continue to work on the problems in Part A without the use of any calculator.

$$\boxed{\text{Section II Part A: Graphing Calculator MAY BE USED.}}$$

1. Let f be the function given by $f(x) = \dfrac{1}{\sqrt{2\pi}}e^{-x^2/2}$ for all real numbers x.

   (a) Find $\displaystyle\lim_{x \to -\infty} f(x)$ and $\displaystyle\lim_{x \to \infty} f(x)$.

   (b) Sketch the graph of f in the window provided.

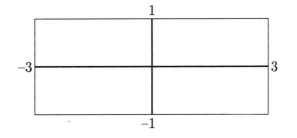

   (c) Find the x-coordinates of the points of inflection.

   (d) Find the area of the region enclosed by the x-axis, the graph of f and the vertical lines containing the points of inflection found in part (c).

   (a) Find $\displaystyle\lim_{x \to -\infty} f(x)$ and $\displaystyle\lim_{x \to \infty} f(x)$.

(b) Sketch the graph of f in the window provided.

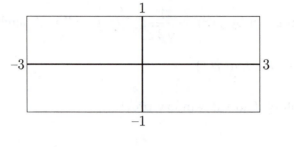

(c) Find the x-coordinates of the points of inflection.

(d) Find the area of the region enclosed by the x-axis, the graph of f and the vertical lines containing the points of inflection found in part (c).

Section II Part A: Graphing Calculator MAY BE USED.

2. Let R be the region enclosed by the graphs of $f(x) = \dfrac{1}{x^2}$, $g(x) = e^{-x}$, and the lines $x = 1$ and $x = k$ where $k > 1$.

(a) Sketch the graphs of f and g on the axes provided below and shade the region R.

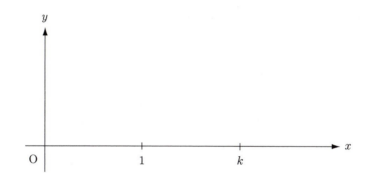

(b) Without using absolute value, set up and evaluate in terms of k, an integral expression that gives $A(k)$, the area of region R.

(c) Find $\displaystyle\lim_{k\to\infty} A(k)$.

(d) Let $k = 4$. Find the volume of the solid generated when region R is revolved around the x-axis.

(a) Sketch the graphs of f and g on the axes provided below and shade the region R.

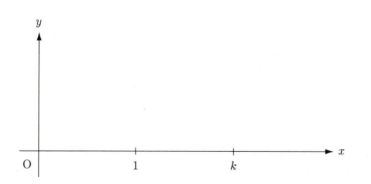

(b)   Without using absolute value, set up and evaluate in terms of k, an integral expression that gives $A(k)$, the area of region R.

(c)   Find $\lim\limits_{k \to \infty} A(k)$.

(d)   Let $k = 4$. Find the volume of the solid generated when region R is revolved around the x-axis.

Section II Part A: Graphing Calculator MAY BE USED.

3.

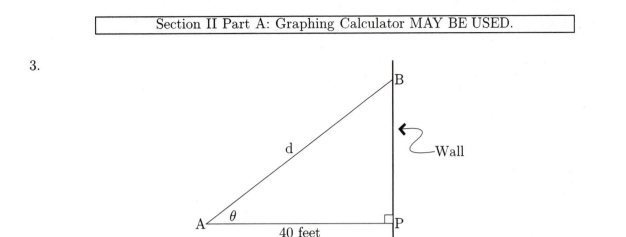

As shown in the figure above, a searchlight is located at point A, 40 feet from a wall. The searchlight revolves counterclockwise at a rate of $\frac{\pi}{30}$ radians per second. At any point B on the wall, the strength of the light, L, is inversely proportional to the square of the distance d from A; that is, at any point on the wall $L = \frac{k}{d^2}$. At the closest point P, $L = 10,000$ lumens.

(a) Find the constant of proportionality k.

(b) Express L as a function of θ, the angle formed by $\overline{AP}$ and $\overline{AB}$.

(c) How fast (in lumens/second) is the strength of the light changing when $\theta = \frac{\pi}{4}$? Is it increasing or decreasing? Justify your answer.

(d) Find the value of θ between $\theta = 0$ and $\theta = \frac{\pi}{2}$ after which $L < 1000$ lumens.

(a) Find the constant of proportionality k.

(b)   Express L as a function of θ, the angle formed by $\overline{AP}$ and $\overline{AB}$.

(c)   How fast (in lumens/second) is the strength of the light changing when $\theta = \frac{\pi}{4}$? Is it increasing or decreasing? Justify your answer.

(d)   Find the value of θ between $\theta = 0$ and $\theta = \frac{\pi}{2}$ after which $L < 1000$ lumens.

Section II Part B: Graphing Calculator MAY **NOT** BE USED.

4. A cylindrical tank is initially filled with water to a depth of 16 feet. A valve in the bottom is opened and the water runs out. The depth, h, of the water in the tank decreases at a rate proportional to the square root of the depth; that is $\frac{dh}{dt} = -k\sqrt{h}$ where k is a constant and $0 < k < 1$.

(a) Find the solution of the differential equation in terms of k.

(b) After the valve is opened, the water falls to a depth of 12.25 feet in 8 hours. Find the value of k.

(c) How many hours after the valve was first opened will the tank be completely empty?

(a) Find the solution of the differential equation in terms of k.

(b)   After the valve is opened, the water falls to a depth of 12.25 feet in 8 hours. Find the value of k.

(c)   How many hours after the valve was first opened will the tank be completely empty?

| Section II Part B: Graphing Calculator MAY **NOT** BE USED. |

5. Consider the curve $y = 4x - x^3$ and chord AB joining points A$(-3, 15)$ and B$(3, -15)$ on the curve.

   (a) Find the x- and y-coordinates of the point(s) on the curve where the tangent line is parallel to chord AB.

   (b) Write an expression without absolute value for the vertical distance, V, between the curve and chord AB for $0 < x < 3$.

   (c) Find the maximum vertical distance between the curve and chord AB for $0 < x < 3$.

   (a) Find the x- and y-coordinates of the point(s) on the curve where the tangent line is parallel to chord AB.

(b)   Write an expression without absolute value for the vertical distance, V, between the curve and chord AB for $0 < x < 3$.

(c)   Find the maximum vertical distance between the curve and chord AB for $0 < x < 3$.

| Section II Part B: Graphing Calculator MAY **NOT** BE USED. |

6. Let the derivative of a function f be $f'(x) = \ln(\sin x)$ and let $f(\frac{\pi}{2}) = 1$.

   (a) Write an equation of the line tangent to f at $x = \frac{\pi}{2}$.

   (b) Write an expression for $f''(x)$.

   (c) Does f change concavity at $x = \frac{\pi}{2}$? Justify your answer.

   (d) Find the Taylor polynomial of degree three centered at $x = \frac{\pi}{2}$, which approximates f.

   (a) Write an equation of the line tangent to f at $x = \frac{\pi}{2}$.

   (b) Write an expression for $f''(x)$.

(c)   Does f change concavity at $x = \frac{\pi}{2}$? Justify your answer.

(d)   Find the Taylor polynomial of degree three centered at $x = \frac{\pi}{2}$, which approximates f.

Sample Examination V

Directions: Solve each of the following problems, using the available space for scratchwork. After examining the form of the choices, decide which is the best of the choices given. Do not spend too much time on any one problem. Calculators may NOT be used on this part of the exam.

In this test: Unless otherwise specified, the domain of a function f is assumed to be the set of all real numbers x for which $f(x)$ is a real number.

1. If $y = (2x^2 + 1)^4$, then $\dfrac{dy}{dx} =$

    (A) $16x^3$

    (B) $4(2x^2 + 1)^3$

    (C) $4x(2x^2 + 1)^3$

    (D) $16(2x^2 + 1)^3$

    (E) $16x(2x^2 + 1)^3$

    Answer

2. $\displaystyle\int x\sqrt{x^2 + 1}\, dx =$

    (A) $\dfrac{x}{\sqrt{x^2 + 1}} + C$

    (B) $\dfrac{3}{4}(x^2 + 1)^{\frac{3}{2}} + C$

    (C) $\dfrac{1}{3}(x^2 + 1)^{\frac{3}{2}} + C$

    (D) $\dfrac{2}{3}(x^2 + 1)^{\frac{3}{2}} + C$

    (E) $\dfrac{1}{3}x^2(x^2 + 1)^{\frac{3}{2}} + C$

    Answer

3. If $f(x) = x^3 - x + 3$ and if c is the only real number such that $f(c) = 0$, then c is between

(A) -2 and -1

(B) -1 and 0

(C) 0 and 1

(D) 1 and 2

(E) 2 and 3

Answer

4. A curve in the plane is defined parametrically by the equations $x = 2t + 3$ and $y = t^2 + 2t$. An equation of the line tangent to the curve at $t = 1$ is

(A) $y = 2x - 7$

(B) $y = x - 2$

(C) $y = 2x$

(D) $y = 2x - 1$

(E) $y = \frac{1}{2}x + \frac{1}{2}$

Answer

5. $\displaystyle\int_0^8 \frac{1}{\sqrt[3]{8 - x}}\, dx$ is

(A) -6 (B) 2 (C) 6 (D) 12 (E) nonexistent

Answer

6. $\int x \sin x \, dx =$

   (A) $-\dfrac{1}{2}x^2 \cos x + C$

   (B) $-x \cos x + C$

   (C) $x \cos x - \sin x + C$

   (D) $-x \cos x + \sin x + C$

   (E) $-x \cos x - \sin x + C$

Answer

---

7. Let f be a differentiable function for all x. Which of the following must be true?

$$\text{I.} \quad \frac{d}{dx} \int_0^3 f(x) \, dx = f(x)$$

$$\text{II.} \quad \int_3^x f'(x) \, dx = f(x)$$

$$\text{III.} \quad \frac{d}{dx} \int_3^x f(x) \, dx = f(x)$$

   (A) II only

   (B) III only

   (C) I and II only

   (D) II and III only

   (E) I, II, and III

Answer

8. If $\sin(xy) = x^2$, then $\dfrac{dy}{dx} =$

(A) $2x \sec(xy)$

(B) $\dfrac{\sec(xy)}{x^2}$

(C) $2x \sec(xy) - y$

(D) $\dfrac{2x \sec(xy)}{y}$

(E) $\dfrac{2x \sec(xy) - y}{x}$

Answer

9. For all x in the closed interval $[1, 4]$, the function g is concave upwards. Which of the following tables could be the values of $g(x)$?

(A)

x	$g(x)$
1	−10
2	−7
3	−6
4	−2

(B)

x	$g(x)$
1	4
2	6
3	9
4	14

(C)

x	$g(x)$
1	0
2	5
3	7
4	12

(D)

x	$g(x)$
1	−4
2	−6
3	−8
4	−10

(E)

x	$g(x)$
1	−2
2	−1
3	5
4	3

Answer

10. $\displaystyle \int \frac{dx}{x^2 + 4x} =$

   (A) $\displaystyle \int \frac{dx}{x} + \int \frac{dx}{x + 4}$

   (B) $\displaystyle \int \frac{dx}{x^2} + \int \frac{dx}{4x}$

   (C) $\displaystyle \int \frac{dx}{x} - \int \frac{dx}{x + 4}$

   (D) $\displaystyle \int \frac{dx}{4x} + \int \frac{dx}{4(x + 4)}$

   (E) $\displaystyle \int \frac{dx}{4x} - \int \frac{dx}{4(x + 4)}$

Answer

11. If $\displaystyle \int_0^4 (x^2 - 6x + 9)\, dx$ is approximated by 4 inscribed rectangles of equal width on the x-axis, then the approximation is

   (A)   14

   (B)   10

   (C)   6

   (D)   5

   (E)   4

Answer

12. What is the 20th derivative of $y = \sin(2x)$?

(A) $-2^{20} \sin(2x)$

(B) $2^{20} \sin(2x)$

(C) $-2^{19} \cos(2x)$

(D) $2^{20} \cos(2x)$

(E) $2^{21} \cos(2x)$

Answer

13. What is an equation of the line tangent to the graph of $f(x) = 7x - x^2$ at the point where $f'(x) = 3$?

(A) $y = 5x - 10$

(B) $y = 3x + 4$

(C) $y = 3x + 8$

(D) $y = 3x - 10$

(E) $y = 3x - 16$

Answer

14. Suppose that $f(x)$ is a twice-differentiable function defined on the closed interval $[a, b]$. If $f'(c) = 0$ for $a < c < b$, which of the following must be true?

I. $f(a) = f(b)$

II. f has a relative extremum at $x = c$.

III. f has a point of inflection at $x = c$.

(A) None

(B) I only

(C) II only

(D) I and II only

(E) II and III only

Answer

15. A sky diver has a negative velocity while falling from an airplane. Before the sky diver opens the parachute, her velocity decreases quickly and then levels off due to air resistance. Which graph approximates the <u>acceleration</u> of the sky diver?

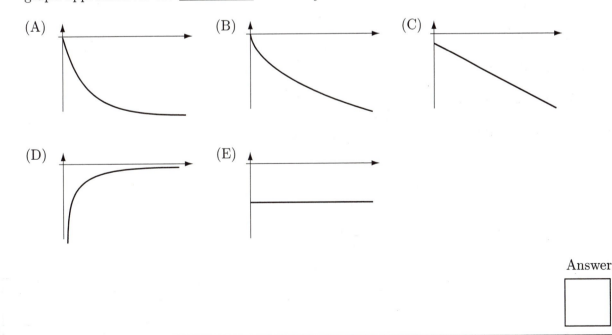

Answer

16. What are the first four nonzero terms in the power series expansion of e^{-4x} about $x = 0$?

(A) $1 + x + \dfrac{x^2}{2} + \dfrac{x^3}{3}$

(B) $1 - 4x + 8x^2 - 32x^3$

(C) $1 - 4x - 2x^2 - \dfrac{2}{3}x^3$

(D) $1 - 4x + 8x^2 - \dfrac{32}{3}x^3$

(E) $1 - 4x + 8x^2 - \dfrac{64}{3}x^3$

Answer

17. If f is a differentiable function such that the slope of the graph of f at each point $(x, f(x))$ is $\sqrt{x^2 - 2x}$, then the length of the graph of f between $(0, f(0))$ and $(2, f(2))$ is

(A) $\dfrac{1}{2}$

(B) $\dfrac{2}{3}$

(C) $\dfrac{3}{4}$

(D) 1

(E) 2

Answer

18. $\displaystyle\int_{e}^{e^{2}} \frac{dx}{x \ln x} =$

(A) $\ln 2$

(B) $\frac{1}{2}$

(C) 1

(D) 2

(E) e

Answer

19. Let $f(x)$ be a continuous and differentiable function on the interval $0 < x < 1$, and let $g(x) = f(3x)$. The table below gives values of $f'(x)$, the derivative of $f(x)$. What is the value of $g'(0.1)$?

x	0.1	0.2	0.3	0.4	0.5	0.6
$f'(x)$	1.01	1.041	1.096	1.179	1.298	1.486

(A) 1.010

(B) 1.096

(C) 1.486

(D) 3.030

(E) 3.288

Answer

20. Which of the following integrals gives the total area of the region shared by both polar curves $r = 2\cos\theta$ and $r = 2\sin\theta$?

(A) $2\displaystyle\int_0^{\frac{\pi}{4}} \sin^2\theta \ d\theta$

(B) $4\displaystyle\int_0^{\frac{\pi}{4}} \sin^2\theta \ d\theta$

(C) $2\displaystyle\int_0^{\frac{\pi}{2}} \sin^2\theta \ d\theta$

(D) $4\displaystyle\int_0^{\frac{\pi}{4}} \cos^2\theta \ d\theta$

(E) $2\displaystyle\int_0^{\frac{\pi}{4}} (\cos^2\theta - \sin^2\theta) \ d\theta$

Answer

21. $\displaystyle\lim_{h\to 0} \frac{2(x+h)^5 - 5(x+h)^3 - 2x^5 + 5x^3}{h}$ is

(A) 0

(B) $10x^3 - 15x$

(C) $10x^4 + 15x^2$

(D) $10x^4 - 15x^2$

(E) $-10x^4 + 15x^2$

Answer

22. If $\int_{2}^{8} f(x)\,dx = -10$ and $\int_{2}^{4} f(x)\,dx = 6$, then $\int_{8}^{4} f(x)\,dx =$

   (A) -16

   (B) -6

   (C) -4

   (D)   4

   (E)   16

Answer

---

23. If the graph of $y = x^3 + ax^2 + bx - 8$ has a point of inflection at $(2, 0)$, what is the value of b?

   (A) 0

   (B) 4

   (C) 8

   (D) 12

   (E) 16

Answer

24. The position of a particle in the xy-plane is given by $x = 4t^2$ and $y = \sqrt{t}$. At $t = 4$, the acceleration vector is

(A) $\left(8, -\frac{1}{64}\right)$ (B) $\left(8, -\frac{1}{32}\right)$ (C) $\left(8, \frac{1}{32}\right)$ (D) $\left(32, -\frac{1}{32}\right)$ (E) $\left(32, \frac{1}{4}\right)$

Answer

25. If f is a continuous function on the closed interval $[a, b]$, which of the following statements are NOT necessarily true?

        I. f has a minimum on $[a, b]$.

        II. f has a maximum on $[a, b]$.

        III. $f'(c) = 0$ for some number c, $a < c < b$.

(A) I only   (B) II only   (C) III only

(D) I and II only   (E) I, II, and III

Answer

26. What are all the values of x for which the series $x - \dfrac{x^2}{2} + \dfrac{x^3}{3} - \dfrac{x^4}{4} + \cdots + (-1)^{n+1}\dfrac{x^n}{n} + \cdots$ converges?

(A) $-1 \le x \le 1$

(B) $-1 \le x < 1$

(C) $-1 < x \le 1$

(D) $-1 < x < 1$

(E) All real numbers

Answer

27. $\displaystyle\sum_{n=0}^{\infty} \frac{(-1)^n (\pi)^{2n}}{(2n)!} =$

(A)   1

(B) -1

(C) π

(D) $\dfrac{\pi}{2}$

(E) e^π

Answer

28. If $\dfrac{dy}{dx} = \dfrac{x}{y}$ and $y(3) = 4$, then

(A) $x^2 - y^2 = -7$

(B) $x^2 + y^2 = 5^2$

(C) $x^2 - y^2 = 7$

(D) $y^2 - x^2 = 5$

(E) $2x^2 - y^2 = 2$

Answer

Section I Part B

Directions: Solve each of the following problems, using the available space for scratchwork. After examining the form of the choices, decide which is the best of the choices given. Do not spend too much time on any one problem. A graphing calculator is required for some questions on this part of the examination.

In this test:

(1) The exact numerical value of the correct answer does not always appear among the choices given. When this happens, select from among the choices the number that best approximates the exact numerical value.

(2) Unless otherwise specified, the domain of a function f is assumed to be the set of all real numbers x for which $f(x)$ is a real number.

29.

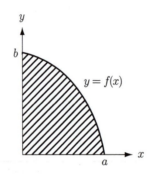

Let $f(x)$ be a continuous function and let A be the area of the shaded region in the figure above. Which of the following statments must be true?

I. $\quad A = \displaystyle\int_0^a f(x)\,dx$

II. $\quad A = \displaystyle\int_0^b f^{-1}(x)\,dx$

III. $\quad A = \displaystyle\int_0^b f^{-1}(y)\,dy$

(A) I only

(B) II only

(C) III only

(D) I and II only

(E) I, II, and III

Answer

30. The Maclaurin series for a function f is given by $\sum_{n=0}^{\infty} \frac{x^n}{2n}$. What is the value of $f^{(4)}(0)$, the fourth derivative of f at $x = 0$?

    (A) 1

    (B) 2

    (C) 3

    (D) 4

    (E) 5

Answer

31. If $f'(x) = (x-a)(x-b)(x-c)$ and $a < b < c$, then which of the following could be the graph of $f(x)$?

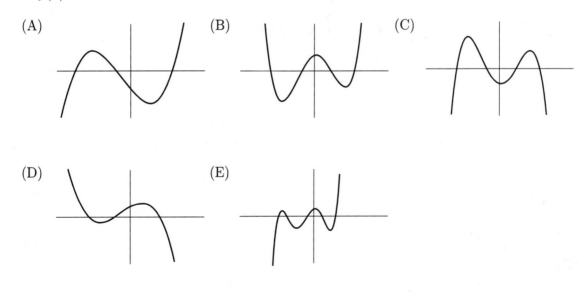

(A)

(B)

(C)

(D)

(E)

Answer

32. If $f(3) = 7$ and $f'(x) = \dfrac{\sin(1 + x^2)}{x^3 - 2x}$, then $f(5) \approx$

(A) -16.006

(B) -9.006

(C) -0.008

(D) 6.992

(E) 7.008

Answer

33. If $\displaystyle\sum_{n=1}^{\infty} |a_n|$ converges, then which of the following is true?

I. $\displaystyle\sum_{n=1}^{\infty} a_n$ converges.

II. $\displaystyle\sum_{n=1}^{\infty} a_n$ is absolutely convergent.

III. $\displaystyle\sum_{n=1}^{\infty} -a_n$ converges.

(A) I only

(B) II only

(C) III only

(D) I and III only

(E) I, II, and III

Answer

34. The base of a solid is the region enclosed by the graph of $y = 3(x - 2)^2$ and the coordinate axes. If every cross section perpendicular to the x-axis is a square, then the volume of the solid is

(A) 8.0

(B) 19.2

(C) 24.0

(D) 25.6

(E) 57.6

Answer

35. When $x = \dfrac{\pi}{4}$, the rate at which $\sin^2 x$ is increasing is k times the rate at which x is increasing. What is the value of k?

(A) $\sqrt{2}$

(B) $\dfrac{\sqrt{2}}{2}$

(C) 1

(D) $\dfrac{1}{2}$

(E) -1

Answer

36. Let f be a function whose seventh derivative is $f^{(7)}(x) = 10,000\cos x$. If $x = 1$ is in the interval of convergence of the power series for this function, then the Taylor polynomial of degree six centered at $x = 0$ will approximate $f(1)$ with an error of not more than

(A) 2.45×10^{-5}

(B) 1.98×10^{-4}

(C) 3.21×10^{-2}

(D) 0.248

(E) 1.984

Answer

---

37. If f is an antiderivative of $\dfrac{\tan^2 x}{x^2 + 1}$ such that $f(1) = \dfrac{1}{2}$, then $f(0) =$

(A) 0

(B) 0.155

(C) 0.345

(D) 0.845

(E) 1

Answer

38. Suppose that $f(x)$, $f'(x)$, and $f''(x)$ are continuous for all real numbers x, and that f has the following properties.

> I. f is negative on $(-\infty, 6)$ and positive on $(6, \infty)$.
>
> II. f is increasing on $(-\infty, 8)$ and decreasing on $(8, \infty)$.
>
> III. f is concave down on $(-\infty, 10)$ and concave up on $(10, \infty)$.

Of the following, which has the <u>smallest</u> numerical value?

(A) $f'(0)$

(B) $f'(6)$

(C) $f''(4)$

(D) $f''(10)$

(E) $f''(12)$

Answer

39. The present average price of a new car is \$14,500. The price of a new car is increasing at a rate of $120 + 180\sqrt{t}$ dollars per year. What will be the approximate average price of a new car five years from now?

(A) \$15,020

(B) \$15,300

(C) \$16,440

(D) \$18,120

(E) \$22,600

Answer

40. If $0 \leq k \leq \frac{\pi}{2}$ and the area of the region in the first quadrant under the graph of $y = 2x - \sin x$ from 0 to k is 0.1, then $k =$

(A) 0.444

(B) 0.623

(C) 0.883

(D) 1.062

(E) 1.571

Answer

41.

x	$f'(x)$
0.998	0.980
0.999	0.995
1.000	1.000
1.001	0.995
1.002	0.980

The table above gives values of the derivative of a function f. Based on this information, it appears that in the interval covered by the table

(A) f is increasing and concave up everywhere.

(B) f is increasing and concave down everywhere.

(C) f has a point of inflection.

(D) f is decreasing and concave up everywhere.

(E) f is decreasing and concave down everywhere.

Answer

42. The mass, $m(t)$, in grams, of a tumor t weeks after it begins growing is given by $m(t) = \dfrac{te^t}{80}$. What is the average rate of change, in grams per week, during the fifth week of growth?

    (A) 2.730

    (B) 3.412

    (C) 6.189

    (D) 6.546

    (E) 11.131

Answer

43.

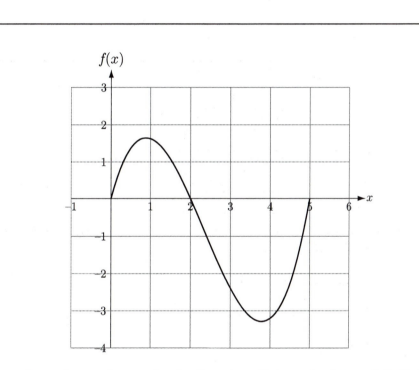

The figure above shows the graph of a function $f(x)$ on the interval $[0, 5]$. Which of the following definite integrals has the greatest value?

(A) $\displaystyle\int_0^1 f(x)\,dx$ (B) $\displaystyle\int_0^2 f(x)\,dx$ (C) $\displaystyle\int_0^3 f(x)\,dx$

(D) $\displaystyle\int_0^4 f(x)\,dx$ (E) $\displaystyle\int_0^5 f(x)\,dx$

Answer

44. The velocity vector of a particle moving in the xy-plane is $(3 - 4\cos t, 4\sin t)$ for all $t \geq 0$. When $t = 0$, the particle is at the point $(0, -1)$. Which statement best describes the motion of the particle?

(A) The particle moves around a circle.

(B) The particle moves along a sine graph.

(C) The particle moves to the left for all t.

(D) The particle moves to the right with a regular up and down motion.

(E) The particle moves generally to the right with a regular up and down motion, but periodically loops to the left.

Answer

45. The closed interval $[0, \pi]$ is partitioned into n equal subdivisions each of length $\Delta x = \frac{\pi}{n}$ by the numbers $x_0, x_1, x_2, \ldots, x_{n-1}, x_n$ with $0 = x_0 < x_1 < x_2 \ldots < x_{n-1} < x_n = \pi$.

$$\lim_{n \to \infty} \sum_{i=1}^{n} x_i \cos(x_i)\Delta x \text{ is}$$

(A) -2

(B) -1

(C) 1

(D) 2

(E) π

Answer

SECTION II - FREE-RESPONSE QUESTIONS
GENERAL INSTRUCTIONS

You may wish to look over the problems before starting to work on them, since it is not expected that everyone will be able to complete all parts of all problems. All problems are given equal weight, but the parts of a particular problem are not necessarily given equal weight.

- YOU SHOULD WRITE ALL WORK FOR EACH PART OF EACH PROBLEM IN THE SPACE PROVIDED FOR THAT PART. Be sure to write clearly and legibly. If you make an error, you may save time by crossing it out rather than trying to erase it. Erased or crossed-out work will not be graded.

- Show all your work. Clearly label any functions, graphs, tables, or other objects that you use. You will be graded on the correctness and completeness of your methods as well as your answers. Answers without supporting work may not receive credit.

- Justifications require that you give mathematical (noncalculator) reasons.

- Your work must be expressed in standard mathematical notation rather than calculator syntax. For example, $\int_1^5 x^2 \, dx$ may not be written as fnInt(X^2, X, 1, 5).

- Unless otherwise specified, answers (numeric or algebraic) need not be simplified.

- If you use decimal approximations in calculations, you will be graded on accuracy. Unless otherwise specified, your final answers should be accurate to three places after the decimal point.

- Unless otherwise specified, the domain of a function f is assumed to be the set of all real numbers x for which $f(x)$ is a real number.

SECTION II PART A: 45 Minutes, Questions 1,2,3

During the timed portion for Part A, you may work only on the problems in Part A. Write your solution to each part of each problem in the space provided for that part.

On Part A, you are permitted to use your calculator to solve an equation, find the derivative of a function at a point, or calculate the value of a definite integral. However, you must clearly indicate the setup of your problem, namely the equation, function, or integral you are using. If you use other built-in features or programs, you must show the mathematical steps necessary to produce your results.

Do not go on to Part B until you are told to do so.

SECTION II PART B: 45 Minutes, Questions 4,5,6

Write your solution to each part of each problem in the space provided for that part. During the timed portion for Part B, you may continue to work on the problems in Part A without the use of any calculator.

Section II Part A: Graphing Calculator MAY BE USED.

1. Consider the parametric function of the form $x(t) = \cos t$ and $y(t) = \sin(nt)$, where n is a positive integer and $0 \leq t \leq 2\pi$.

    (a) On the axes provided, sketch the graphs for $n = 3$ and $n = 4$.

    (b) Give an expression for $\dfrac{dy}{dx}$ in terms of t and n.

    (c) In terms of n, at how many points will the tangent line to the graph be horizontal? Use $\dfrac{dy}{dx}$ to justify your answer.

    (d) For what values of t will the tangent line be vertical? Justify your answer.

(a) On the axes provided, sketch the graphs for $n = 3$ and $n = 4$.

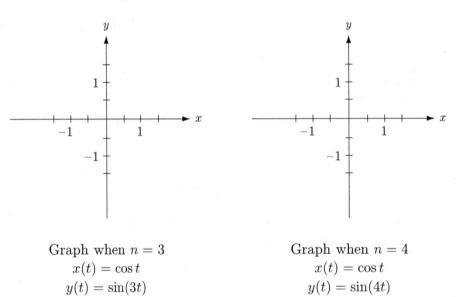

Graph when $n = 3$

$x(t) = \cos t$

$y(t) = \sin(3t)$

Graph when $n = 4$

$x(t) = \cos t$

$y(t) = \sin(4t)$

(b) Give an expression for $\dfrac{dy}{dx}$ in terms of t and n.

(c) In terms of n, at how many points will the tangent line to the graph be horizontal? Use $\dfrac{dy}{dx}$ to justify your answer.

(d) For what values of t will the tangent line be vertical? Justify your answer.

> Section II Part A: Graphing Calculator MAY BE USED.

2. Let $f(x)$ be a function with the following properties.

> i. $f(0) = 1$
>
> ii.   For all integers $n \geq 0$, the nth derivative, $f^{(n)}(x) = (-1)^n a^n f(x)$, where $a > 0$ and $a \neq 1$.

(a) Write the first four non-zero terms of the power series of $f(x)$ centered at zero, in terms of a.

(b) Write $f(x)$ as a familiar function in terms of a.

(c) How many terms of the power series are necessary to approximate $f(0.2)$ with an error less than 0.001 with $a = 2$?

(a) Write the first four non-zero terms of the power series of $f(x)$ centered at zero, in terms of a.

(b) Write $f(x)$ as a familiar function in terms of a.

(c) How many terms of the power series are necessary to approximate $f(0.2)$ with an error less than 0.001 with $a = 2$?

Section II Part A: Graphing Calculator MAY BE USED.

3.

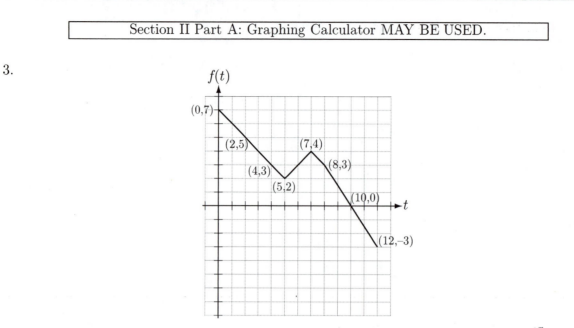

The figure above shows the graph of a continuous function f. Let $s(x) = \int_0^x f(t)\,dt$ for $0 \le x \le 12$.

(a)   What feature of the graph does $s(x)$ measure?

(b)   Find the value of $s(4)$ and $s'(4)$.

(c)   When, if ever, does $s'(x) = 0$? Why?

(d)   For what value of x does $s(x)$ attain its maximum value? Explain your answer.

(a)   What feature of the graph does $s(x)$ measure?

(b) Find the value of $s(4)$ and $s'(4)$.

(c) When, if ever, does $s'(x) = 0$? Why?

(d) For what value of x does $s(x)$ attain its maximum value? Explain your answer.

Section II Part B: Graphing Calculator MAY **NOT** BE USED.

4.

hours	0	0.1	0.2	0.3	0.4	0.5	0.6	0.7	0.8	0.9	1.0
a m/h/h	90	90	90	80	78	80	60	40	30	10	0

A train, initially stopped, begins moving. The table above shows the train's acceleration, a, in miles/hour/hour as a function of time measured in hours.

Let $f(t) = \displaystyle\int_0^t a(x)\, dx$

(a)   Use the midpoint Riemann sum with 5 subdivisions of equal length to approximate $f(1)$.

(b)   Explain what the value of $f(1)$ represents, and give its units of measure.

(c)   Assume that the train's acceleration is constant on the interval $[0, 0.2]$. How far does the train travel during this interval? Include units of measure. Show how you arrived at your answer.

(a)   Use the midpoint Riemann sum with 5 subdivisions of equal length to approximate $f(1)$.

(b) Explain what the value of $f(1)$ represents, and give its units of measure.

(c) Assume that the train's acceleration is constant on the interval $[0, 0.2]$. How far does the train travel during this interval? Include units of measure. Show how you arrived at your answer.

5.

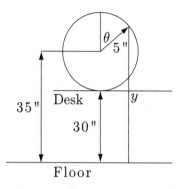

As shown in the figure above, a clock sits on a desk so that its center is 35 inches from the floor. The *minute hand* is 5 inches long. Let θ be the angle between the 12 o'clock position and the minute's hand position at any time t minutes after the hour. (Note: θ is measured clockwise).

(a)   Using a trigonometric function of θ, write an expression for y, the distance from the tip of the minute hand to the floor.

(b)   Express θ as a function of t.

(c)   In terms of θ, find an expression for $\dfrac{dy}{dt}$, the rate at which y changes.

(d)   At how many minutes after the hour is y increasing most rapidly? Use $\dfrac{dy}{dt}$ to justify your answer.

(a)   Using a trigonometric function of θ, write an expression for y, the distance from the tip of the minute hand to the floor.

(b)   Express θ as a function of t.

(c)   In terms of θ, find an expression for $\dfrac{dy}{dt}$, the rate at which y changes.

(d)   At how many minutes after the hour is y increasing most rapidly? Use $\dfrac{dy}{dt}$ to justify your answer.

Section II Part B: Graphing Calculator MAY **NOT** BE USED.

---

6. The differential equation $14 = \dfrac{dT}{dt} + 0.01T$ models the temperature, T, above room temperature, of an electric burner. T is measured in degrees Fahrenheit ($^\circ$F) and $t \geq 0$ is in seconds.

(a) Find the general solution of the differential equation in terms of a constant k.

(b) When the burner is first turned on at room temperature, $T = 0^\circ$F. Use this fact to find the value of the constant k.

(c) The Safety Standards Board has determined that this type of burner is unsafe if its temperature can exceed 1500°F. Is this burner safe or unsafe? Justify your answer.

(a) Find the general solution of the differential equation in terms of a constant k.

(b) When the burner is first turned on at room temperature, $T = 0°\text{F}$. Use this fact to find the value of the constant k.

(c) The Safety Standards Board has determined that this type of burner is unsafe if its temperature can exceed $1500°\text{F}$. Is this burner safe or unsafe? Justify your answer.

Sample Examination VI

Directions: Solve each of the following problems, using the available space for scratchwork. After examining the form of the choices, decide which is the best of the choices given. Do not spend too much time on any one problem. Calculators may NOT be used on this part of the exam.

In this test: Unless otherwise specified, the domain of a function f is assumed to be the set of all real numbers x for which $f(x)$ is a real number.

1. The slope field for the differential equation $\dfrac{dy}{dx} = \dfrac{x^2 y + y^2 x}{3 + y}$ will have horizontal segments when

   (A) $x = 0$ or $y = 0$, only

   (B) $y = -x$, only

   (C) $y = -3$, only

   (D) $y = 5$, only

   (E) $x = 0$, or $y = 0$, or $y = -x$

   Answer

---

2. $\displaystyle\int_0^2 xe^x \, dx =$

   (A) $1 - e^2$

   (B) $e^2 - 1$

   (C) $e^2 + 1$

   (D) $e^4 - e^2 + 1$

   (E) $e^4 + e^2 - 1$

   Answer

3. A particle moves in the xy-plane so that at any time t its coordinates are $x = \alpha \cos \beta t$ and $y = \alpha \sin \beta t$, where α and β are constants. The y-component of the acceleration of the particle at any time t is

    (A) $-\beta^2 y$

    (B) $-\beta^2 x$

    (C) $-\alpha\beta \sin \beta t$

    (D) $-\alpha\beta \cos \beta t$

    (E) $-\beta^2 \alpha \cos \beta t$

Answer

4. An equation of the tangent line to the curve $y = \dfrac{3x + 4}{4x - 3}$ at the point $(1, 7)$ is

    (A) $y = -25x + 32$

    (B) $y = 31x - 24$

    (C) $y = 7x$

    (D) $y = -5x + 12$

    (E) $y = 25x - 18$

Answer

5. A particle moves along the x-axis so that at any time t its position is given by $x(t) = \frac{1}{2}\sin(t) + \cos(2t)$. What is the acceleration of the particle at $t = \frac{\pi}{2}$?

(A) 0

(B) $\frac{1}{2}$

(C) $\frac{3}{2}$

(D) $\frac{5}{2}$

(E) $\frac{7}{2}$

Answer

6.

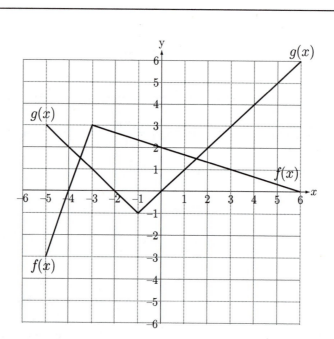

The functions f and g are piecewise linear functions whose graphs are shown above. If $h(x) = f(g(x))$, then $h'(-3) =$

(A) -3 (B) $-\frac{1}{3}$ (C) 0

(D) $\frac{1}{3}$ (E) 3

Answer

7. Let the repeating decimal $0.242424\ldots = N$. Which statement is true?

$$\text{I.} \quad N = \sum_{k=1}^{\infty} 24\left(\frac{1}{100}\right)^k$$

$$\text{II.} \quad N = \frac{24}{1 - 100^{-2}}$$

$$\text{III.} \quad N = \frac{8}{33}$$

(A) I only

(B) II only

(C) III only

(D) I and II only

(E) I and III only

Answer

8. $\displaystyle \int_0^3 \frac{x}{\sqrt{x^2 + 16}}\, dx =$

(A) 1

(B) 2

(C) 3

(D) 4

(E) 5

Answer

9. If $y = \ln(3x + 5)$, then $\dfrac{d^2y}{dx^2} =$

(A) $\dfrac{3}{3x + 5}$

(B) $\dfrac{3}{(3x + 5)^2}$

(C) $\dfrac{9}{(3x + 5)^2}$

(D) $\dfrac{-9}{(3x + 5)^2}$

(E) $\dfrac{-3}{(3x + 5)^2}$

Answer

10. $\displaystyle\int_{-1}^{1} \dfrac{dx}{x^2 + 5x + 6} =$

(A) $\ln \dfrac{3}{2}$

(B) $\ln \dfrac{1}{4}$

(C) $\ln \dfrac{2}{3}$

(D) $\ln 6$

(E) $\ln 12$

Answer

11. If $y^2 - 2xy = 21$, then $\dfrac{dy}{dx}$ at the point $(2, -3)$ is

(A) $-\dfrac{6}{5}$

(B) $-\dfrac{3}{5}$

(C) $-\dfrac{2}{5}$

(D) $\dfrac{3}{8}$

(E) $\dfrac{3}{5}$

Answer

12. The average value of $\sqrt{3x}$ on the closed interval $[0, 9]$ is

(A) $\dfrac{2\sqrt{3}}{3}$

(B) $2\sqrt{3}$

(C) 6

(D) $6\sqrt{3}$

(E) $18\sqrt{3}$

Answer

13. If $C(x)$ gives the cost in dollars of producing x items of a certain product, which of the following statements are true about $\dfrac{dC}{dx}$, the derivative of $C(x)$?

    I.   The units of $\dfrac{dC}{dx}$ are dollars per item.

    II.   The value of $\dfrac{dC}{dx}$ at any value of x is the cost of producing one additional item.

    III. $\dfrac{dC}{dx}$ is the rate at which items are produced.

(A) I only

(B) II only

(C) III only

(D) I and II only

(E) I, II, and III

Answer

14. $\displaystyle\int_{-\infty}^{\infty} e^{-|x|}\, dx$ is

(A)   0

(B) −1

(C)   1

(D)   2

(E)   nonexistent

Answer

15. On which of the following intervals is the graph of $y = 6x^2 + \frac{x}{2} + 3 + \frac{6}{x}$ concave downward?

   (A) $x < -1$

   (B) $x < 0$

   (C) $-1 < x < 0$

   (D) $0 < x < 1$

   (E) $x > -1$

Answer

16. What is the area of the largest rectangle with lower base on the x-axis and upper vertices on the curve $y = 12 - x^2$?

   (A)   8

   (B)   12

   (C)   16

   (D)   32

   (E)   48

Answer

17. The efficiency of an automobile engine is given by the continuous function $r(c)$ where r is measured in liters/kilometer and c is measured in kilometers. What are the units of $\int_0^5 r(c) \, dc$?

(A) liters

(B) kilometers

(C) liter-kilometers

(D) liters/kilometer

(E) kilometers/liter

Answer

18.

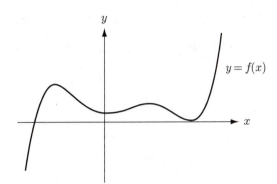

If the graph of a fifth-degree polynomial, $f(x)$, is shown above, then the graph of $f'(x)$, the derivative of $f(x)$, will cross the x-axis in exactly how many points?

(A) None

(B) One

(C) Two

(D) Three

(E) Four

Answer

19. At what point on the curve $x^2 - y^2 + x = 2$ is the tangent line vertical?

(A) $(1, 0)$ only

(B) $(-2, 0)$ only

(C) $(1, \sqrt{2})$ only

(D) $(1, 0)$ and $(-2, 0)$

(E) The tangent line is never vertical.

Answer

20.

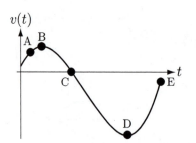

The figure above shows the graph of the velocity of an object moving on the x-axis as a function of time. Which of the marked points corresponds to the time when the object is farthest to the right?

(A) A

(B) B

(C) C

(D) D

(E) E

Answer

21. If $\lim\limits_{x \to 2} \dfrac{f(x)}{x - 2} = f'(2) = 0$, which of the following must be true?

      I. $f(2) = 0$

      II. $f(x)$ is continuous at $x = 2$.

      III. $f(x)$ has a horizontal tangent line at $x = 2$.

(A) I only

(B) II only

(C) I and II only

(D) II and III only

(E) I, II, and III

Answer

22. $\int_1^\infty xe^{-x}\,dx$ is

(A) 0

(B) $\dfrac{1}{e}$

(C) $\dfrac{2}{e}$

(D) 1

(E) divergent

Answer

23. If $f(x) = \dfrac{\sin^2 x}{1 - \cos x}$, then $f'(x) =$

(A) $\cos x$

(B) $\sin x$

(C) $-\sin x$

(D) $-\cos x$

(E) $2\cos x$

Answer

24. If $\dfrac{dy}{dx} = y \cos x$ and $y = 3$ when $x = 0$, then $y =$

(A) $e^{\sin x} + 2$

(B) $e^{\sin x} + 3$

(C) $\sin x + 3$

(D) $\sin x + 3e^x$

(E) $3e^{\sin x}$

Answer

---

25. Lef $f(x)$ be a differentiable function with no points of inflection on $[a, b]$. If the definite integral $\int_a^b f(x)\, dx > T$, where T is the Trapezoidal Rule approximation to $\int_a^b f(x)\, dx$, which of the following statements about $f(x)$ must be true?

(A) $f(x)$ is linear.

(B) $f(x)$ is concave upwards on $[a, b]$.

(C) $f(x)$ is concave downwards on $[a, b]$.

(D) $f(x)$ is increasing on $[a, b]$.

(E) $f(x)$ is decreasing on $[a, b]$.

Answer

26. How many extrema (maxima and minima) does the function $f(x) = (x+2)^3(x-5)^2$ have?

    (A) None

    (B) One

    (C) Two

    (D) Three

    (E) Four

Answer

27. $\displaystyle\lim_{n\to\infty} \sum_{k=1}^{n} \left(2 + \frac{3}{n}k\right)^2 \left(\frac{3}{n}\right) =$

    (A)   0

    (B)   1

    (C)   4

    (D)   39

    (E)   125

Answer

28.

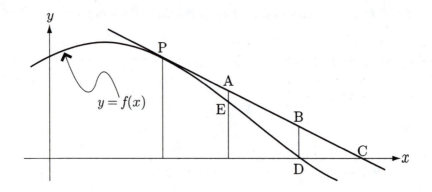

In the figure above $\overleftrightarrow{PC}$ is tangent to the graph of $y = f(x)$ at point P. Points A and B are on $\overleftrightarrow{PC}$ and points D and E are on the graph of $y = f(x)$. Which of the following statements are true?

I.   Euler's method uses the y-coordinate of point A to approximate the y-coordinate at point E.

II.   Euler's method uses the x-coordinate of point B to approximate the root of the function $f(x)$ at point D.

III.   Euler's method uses the x-coordinate of point C to approximate a root of the function $f(x)$ at point D.

(A) I only

(B) II only

(C) III only

(D) I and II only

(E) I and III only

Answer

Section I Part B

Directions: Solve each of the following problems, using the available space for scratchwork. After examining the form of the choices, decide which is the best of the choices given. Do not spend too much time on any one problem. A graphing calculator is required for some questions on this part of the examination.

In this test:

(1) The exact numerical value of the correct answer does not always appear among the choices given. When this happens, select from among the choices the number that best approximates the exact numerical value.

(2) Unless otherwise specified, the domain of a function f is assumed to be the set of all real numbers x for which $f(x)$ is a real number.

29. Let f be a continuous function such that $\int_{2}^{3} f(2x)\, dx = 8$. What is the value of $\int_{4}^{6} f(x)\, dx$?

(A) 4

(B) 8

(C) 12

(D) 16

(E) 32

Answer

30. The area enclosed by the polar curve $r = 6\cos\theta + 8\sin\theta$ from $\theta = 0$ to $\theta = \pi$ is

(A) 28.274

(B) 50.265

(C) 78.540

(D) 113.097

(E) 201.062

Answer

<u>Questions 31 and 32</u> refer to the following information.

The roof and walls of a storage building are built in the shape modeled by the curve $y(x) = 20 - \dfrac{x^6}{3,200,000}$. Each cross section cut perpendicular to the x-axis is a rectangle with a base of 50 feet and a height of y feet.

31. In cubic feet the volume of the building is approximately

(A) 686

(B) 2,000

(C) 17,100

(D) 34,300

(E) 50,000

Answer

32. What is the average height in feet of the storage building described above?

(A) 8.571

(B) 11.920

(C) 12.500

(D) 15.071

(E) 17.143

Answer

33. A company manufactures x calculators weekly that can be sold for $75 - 0.01x$ dollars each. The cost of manufacturing x calculators is given by $1850 + 28x - x^2 + 0.001x^3$. The number of calculators the company should manufacture weekly in order to maximize its weekly profit is

(A) 611

(B) 652

(C) 683

(D) 749

(E) 754

Answer

34.

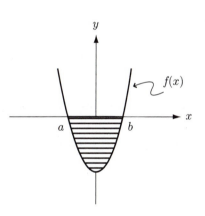

If f is the continuous function shown in the figure above, then the area of the shaded region is

(A) $\displaystyle\int_a^b f(x)\,dx$

(B) $\displaystyle\int_b^a f(x)\,dx$

(C) $\displaystyle\int_b^{-a} f(x)\,dx$

(D) $\displaystyle\int_{-a}^b f(x)\,dx$

(E) $\displaystyle\int_{-b}^{-a} f(x)\,dx$

Answer

35. A missile rises vertically from a point on the ground 75,000 feet from a radar station. If the missile is rising at the rate of 16,500 feet per minute at the instant when it is 38,000 feet high, what is the rate of change, in radians per minute, of the missile's angle of elevation from the radar station at this instant?

(A) 0.175

(B) 0.219

(C) 0.227

(D) 0.469

(E) 0.507

Answer

36. Let $f(x)$ be a function whose Taylor series converges for all x. If $|f^{(n)}(x)| < 1$ where $f^{(n)}(x)$ is the nth derivative of $f(x)$, what is the minimum number of terms of the Taylor series centered at $x = 1$ necessary to approximate $f(1.2)$ with an error less than 0.00001?

(A) Three

(B) Four

(C) Five

(D) Six

(E) Ten

Answer

37. What are all values of p for which $\displaystyle\int_1^\infty \frac{1}{x^{\pi p}}\, dx$ converges?

(A) $p > 0$

(B) $p > \dfrac{1}{\pi}$

(C) $p > 1$

(D) $p > \pi$

(E) There is no value of p for which the integral converges.

Answer

38.

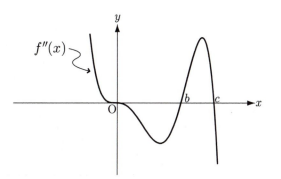

The figure above shows the graph of $f''(x)$, the second derivative of a function $f(x)$. The function $f(x)$ is continuous for all x. Which of the following statements about f are true?

     I. f is concave up for $x < 0$ and $b < x < c$.

     II. f has a relative minimum in the open interval $b < x < c$.

     III. f has points of inflection at $x = 0$ and $x = b$.

(A) I only

(B) II only

(C) III only

(D) I and III only

(E) I, II, and III

Answer

39. If $f(x) > 0$ is continuous and $g(x) = \int_0^x \sqrt{(f(t))^2 - 1}\, dt$, what is the length of the graph of $g(x)$ from $x = a$ to $x = b$?

(A) $\int_a^b f(x)\, dx$

(B) $\int_a^b g(x)\, dx$

(C) $\int_a^b \sqrt{(f(x))^2 + 1}\, dx$

(D) $\int_a^b \sqrt{g(x) + 1}\, dx$

(E) $\int_a^b \sqrt{(g(x))^2 + 1}\, dx$

Answer

40. Let $T(x) = \sum_{k=0}^{\infty} \left(\frac{1}{2}\right)^k \frac{(x-3)^k}{k!}$ be the Taylor series for a function f.

What is the value of $f^{(10)}(3)$, the tenth derivative of f at $x = 3$?

(A) 5.382×10^{-10}

(B) 2.691×10^{-10}

(C) 9.766×10^{-4}

(D) 4.883×10^{-4}

(E) 1.953×10^{-3}

Answer

41. $\dfrac{d}{dx} \displaystyle\int_0^{2x} (e^t + 2t)\, dt =$

   (A) $e^{2x} + 4x$

   (B) $e^{2x} + 4x - 1$

   (C) $e^{2x} + 4x^2 - 1$

   (D) $2e^{2x} + 4x$

   (E) $2e^{2x} + 8x$

Answer

42. On the interval $[0, b]$ the number $c = 4.522$ is the number guaranteed by the Mean Value Theorem for the function $f(x) = \sin(x)$. What is the approximate value of b?

   (A) 3

   (B) 3.25

   (C) 3.5

   (D) 3.75

   (E) 4

Answer

43. What is the approximate value of $\cos(\frac{1}{2})$ obtained by using a fourth-degree Taylor Polynomial for $\cos x$ centered at $x = 0$?

(A) $\frac{1}{2} - \frac{1}{48} + \frac{1}{3840}$

(B) $\frac{1}{2} - \frac{1}{24} + \frac{1}{640}$

(C) $1 - \frac{1}{4} + \frac{1}{16}$

(D) $1 - \frac{1}{8} + \frac{1}{64}$

(E) $1 - \frac{1}{8} + \frac{1}{384}$

Answer

44. If $\displaystyle\int_0^{1000} 8^x \; dx \; - \int_a^{1000} 8^x \; dx = 10.40$, then a is approximately

(A) 1.4

(B) 1.5

(C) 1.6

(D) 1.7

(E) 1.8

Answer

45.

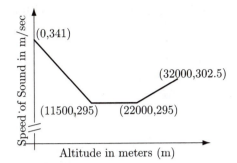

In the earth's atmosphere the speed of sound is a function of the altitude. The figure above, consisting of 3 line segments, shows the speed of sound, $s(a)$, in m/sec as a function of altitude, a, in meters. The graph is not drawn to scale. What is the average speed of sound in m/sec on the interval $[0, 32000]$?

(A) 295

(B) 303.9

(C) 304.4

(D) 306.8

(E) 312.8

Answer

SECTION II - FREE-RESPONSE QUESTIONS
GENERAL INSTRUCTIONS

You may wish to look over the problems before starting to work on them, since it is not expected that everyone will be able to complete all parts of all problems. All problems are given equal weight, but the parts of a particular problem are not necessarily given equal weight.

- YOU SHOULD WRITE ALL WORK FOR EACH PART OF EACH PROBLEM IN THE SPACE PROVIDED FOR THAT PART. Be sure to write clearly and legibly. If you make an error, you may save time by crossing it out rather than trying to erase it. Erased or crossed-out work will not be graded.

- Show all your work. Clearly label any functions, graphs, tables, or other objects that you use. You will be graded on the correctness and completeness of your methods as well as your answers. Answers without supporting work may not receive credit.

- Justifications require that you give mathematical (noncalculator) reasons.

- Your work must be expressed in standard mathematical notation rather than calculator syntax. For example, $\int_{1}^{5} x^2 \, dx$ may not be written as fnInt(X^2, X, 1, 5).

- Unless otherwise specified, answers (numeric or algebraic) need not be simplified.

- If you use decimal approximations in calculations, you will be graded on accuracy. Unless otherwise specified, your final answers should be accurate to three places after the decimal point.

- Unless otherwise specified, the domain of a function f is assumed to be the set of all real numbers x for which $f(x)$ is a real number.

SECTION II PART A: 45 Minutes, Questions 1,2,3

During the timed portion for Part A, you may work only on the problems in Part A. Write your solution to each part of each problem in the space provided for that part.

On Part A, you are permitted to use your calculator to solve an equation, find the derivative of a function at a point, or calculate the value of a definite integral. However, you must clearly indicate the setup of your problem, namely the equation, function, or integral you are using. If you use other built-in features or programs, you must show the mathematical steps necessary to produce your results.

Do not go on to Part B until you are told to do so.

SECTION II PART B: 45 Minutes, Questions 4,5,6

Write your solution to each part of each problem in the space provided for that part. During the timed portion for Part B, you may continue to work on the problems in Part A without the use of any calculator.

Section II Part A: Graphing Calculator MAY BE USED.

1. Two particles move in the xy-plane: particle A moves along the graph of $y = 1 + e^{-x}$ and particle B moves along the graph of $y = \cos x$ for $0 \le x \le 3\pi$. At all times their x-coordinates are the same.

   (a) Sketch the graph of the paths of the particles on the axes provided and label each.

   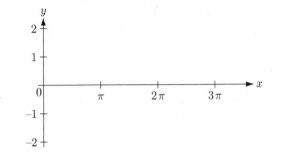

   (b) Do the particles ever collide? Justify your answer.

   (c) What is the minimum vertical distance between the particles? (Show the computations that lead to your answer.)

   (d) What is the maximum vertical distance between the particles? (Show the computations that lead to your answer.)

   (a) Sketch the graph of the paths of the particles on the axes provided and label each.

   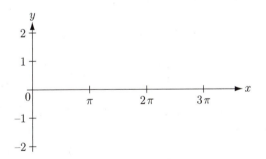

(b)   Do the particles ever collide? Justify your answer.

(c)   What is the minimum vertical distance between the particles? (Show the computations that lead to your answer.)

(d)   What is the maximum vertical distance between the particles? (Show the computations that lead to your answer.)

Section II Part A: Graphing Calculator MAY BE USED.

2. The following table gives the velocity, v, in the vertical direction (in ft/sec) of a rider on a ferris wheel at an amusement park. The time, t, is measured in seconds after the ride starts. The rider moves smoothly and the table gives the values for one complete revolution of the wheel.

t	v
seconds	feet/second
0	0
5	1.6
10	2.7
15	3.1
20	2.7
25	1.6
30	0
35	−1.6
40	−2.7
45	−3.1
50	−2.7
55	−1.6
60	0

(a) During what interval of time is the acceleration negative? Give a reason for your answer.

(b) What is the average acceleration during the first 15 seconds of the ride? Include units of measure.

(c) Approximate $\int_0^{30} v(t)\, dt$ using a Riemann sum with six intervals of equal length.

(d) Approximate the <u>diameter</u> of the ferris wheel. Explain your reasoning.

(a) During what interval of time is the acceleration negative? Give a reason for your answer.

(b) What is the average acceleration during the first 15 seconds of the ride? Include units of measure.

(c) Approximate $\int_0^{30} v(t)\, dt$ using a Riemann sum with six intervals of equal length.

(d) Approximate the <u>diameter</u> of the ferris wheel. Explain your reasoning.

3.

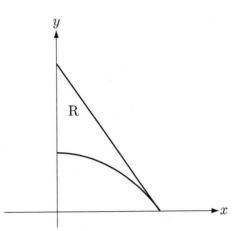

As shown in the accompanying diagram, the region R lies in the first quadrant above the graph of $f(x) = 4 - x^2$ and below the line $y = m(x - 2)$.

(a) If in the first quadrant the line lies above the graph of f, determine the range of m.

(b) When the line intersects the y-axis at $(0, 12)$, what is the area of R?

(c) Write an expression without an integral sign for $A(m)$, the area of R in terms of m.

(d) If m is changing at the constant rate of -2 units per second, how fast is $A(m)$ changing at the instant the line intersects the axis at $(0, 12)$? Is the area increasing or decreasing?

(a) If in the first quadrant the line lies above the graph of f, determine the range of m.

(b)   When the line intersects the y-axis at $(0, 12)$, what is the area of R?

(c)   Write an expression without an integral sign for $A(m)$, the area of R in terms of m.

(d)   If m is changing at the constant rate of -2 units per second, how fast is $A(m)$ changing at the instant the line intersects the axis at $(0, 12)$? Is the area increasing or decreasing?

> Section II Part B: Graphing Calculator MAY **NOT** BE USED.

4. A particle is moving in the xy-plane so that for all t:

$$\frac{dx}{dt} = 4\cos t \text{ and } \frac{dy}{dt} = \sin t$$

At $t = 0$, the particle is at the origin.

(a) Find $x(t)$ and $y(t)$, the position of the particle.

(b) Find the average rate of change of y with respect to x as t varies from 0 to $\frac{\pi}{2}$.

(c) At what time t, $0 \leq t \leq \frac{\pi}{2}$, will the instantaneous rate of change of y with respect to x be equal to the average rate of change of y with respect to x?

(a) Find $x(t)$ and $y(t)$, the position of the particle.

(b)   Find the average rate of change of y with respect to x as t varies from 0 to $\frac{\pi}{2}$.

(c)   At what time t, $0 \le t \le \frac{\pi}{2}$, will the instantaneous rate of change of y with respect to x be equal to the average rate of change of y with respect to x?

5. The fifth-degree Maclaurin polynomial for $\tan(x)$ is $M(x) = x + \frac{1}{3}x^3 + \frac{2}{15}x^5$.

    (a)   Use $M(x)$ to find the 4ᵗʰ degree Maclaurin polynomial for $\sec^2(x)$.

    (b)   Use $M(x)$ to find the 4ᵗʰ degree Maclaurin polynomial for $\dfrac{\tan(x)}{x}$.

    (c)   Use the polynomial found in (b) to find $\displaystyle\lim_{x \to 0} \dfrac{\tan(x)}{x}$.

    (d)   Is the limit found in (c) exact or an approximation? Explain your reasoning.

    (a)   Use $M(x)$ to find the 4ᵗʰ degree Maclaurin polynomial for $\sec^2(x)$.

    (b)   Use $M(x)$ to find the 4ᵗʰ degree Maclaurin polynomial for $\dfrac{\tan(x)}{x}$.

(c)   Use the polynomial found in (b) to find $\lim\limits_{x \to 0} \dfrac{\tan(x)}{x}$.

(d)   Is the limit found in (c) exact or an approximation? Explain your reasoning.

---

6. Consider the differential equation $y + \dfrac{dy}{dx} = e^{-x}$

    (a) The slope field for the differential equation is provided. Sketch the solution curve that passes through $(0,0)$, and sketch the solution curve that passes through $(-2,0)$.

    (b) For any constant C verify that $y = xe^{-x} + Ce^{-x}$ is a solution to the given differential equation.

    (c) Each solution curve has a maximum value and no minimum value. In terms of C find the coordinates of the point where the maximum value occurs and find a function, $f(x)$, on which all the maximum points lie. Show the work that leads to your answer.

    (a) The slope field for the differential equation is provided. Sketch the solution curve that passes through $(0,0)$, and sketch the solution curve that passes through $(-2,0)$.

(b) For any constant C verify that $y = xe^{-x} + Ce^{-x}$ is a solution to the given differential equation.

(c) Each solution curve has a maximum value and no minimum value. In terms of C find the coordinates of the point where the maximum value occurs and find a function, $f(x)$, on which all the maximum points lie. Show the work that leads to your answer.

Formulas and Theorems for Reference

I. Differentiation Formulas

1. $\dfrac{d}{dx}(x^n) = nx^{n-1}$

2. $\dfrac{d}{dx}(fg) = fg' + gf'$

3. $\dfrac{d}{dx}\left(\dfrac{f}{g}\right) = \dfrac{gf' - fg'}{g^2}$

4. $\dfrac{d}{dx}f(g(x)) = f'(g(x))g'(x)$

5. $\dfrac{d}{dx}(\sin x) = \cos x$

6. $\dfrac{d}{dx}(\cos x) = -\sin x$

7. $\dfrac{d}{dx}(\tan x) = \sec^2 x$

8. $\dfrac{d}{dx}(\cot x) = -\csc^2 x$

9. $\dfrac{d}{dx}(\sec x) = \sec x \tan x$

10. $\dfrac{d}{dx}(\csc x) = -\csc x \cot x$

11. $\dfrac{d}{dx}(e^x) = e^x$

12. $\dfrac{d}{dx}(a^x) = a^x \ln a$

13. $\dfrac{d}{dx}(\ln x) = \dfrac{1}{x}$

14. $\dfrac{d}{dx}(\text{Arcsin } x) = \dfrac{1}{\sqrt{1 - x^2}}$

15. $\dfrac{d}{dx}(\text{Arctan } x) = \dfrac{1}{1 + x^2}$

II. Integration Formulas

1. $\displaystyle\int a\,dx = ax + C$

2. $\displaystyle\int x^n\,dx = \dfrac{x^{n+1}}{n+1} + C, n \neq -1$

3. $\displaystyle\int \dfrac{1}{x}\,dx = \ln|x| + C$

4. $\displaystyle\int e^x\,dx = e^x + C$

5. $\displaystyle\int a^x\,dx = \dfrac{a^x}{\ln a} + C$

6. $\displaystyle\int \ln x\,dx = x \ln x - x + C$

7. $\displaystyle\int \sin x\,dx = -\cos x + C$

8. $\displaystyle\int \cos x\,dx = \sin x + C$

9. $\displaystyle\int \tan x\,dx = \ln|\sec x| + C \text{ or } -\ln|\cos x| + C$

10. $\displaystyle\int \cot x\,dx = \ln|\sin x| + C$

11. $\displaystyle\int \sec x\,dx = \ln|\sec x + \tan x| + C$

12. $\displaystyle\int \csc x\,dx = \ln|\csc x - \cot x| + C$

13. $\displaystyle\int \sec^2 x\,dx = \tan x + C$

14. $\displaystyle\int \sec x \tan x\,dx = \sec x + C$

15. $\displaystyle\int \csc^2 x\,dx = -\cot x + C$

16. $\displaystyle\int \csc x \cot x\,dx = -\csc x + C$

17. $\displaystyle\int \dfrac{dx}{a^2 + x^2} = \dfrac{1}{a}\text{Arctan}\left(\dfrac{x}{a}\right) + C$

18. $\displaystyle\int \dfrac{dx}{\sqrt{a^2 - x^2}} = \text{Arcsin}\left(\dfrac{x}{a}\right) + C$

213

III. Formulas and Theorems

1. <u>Limits and Continuity</u>

    A function $y = f(x)$ is continuous at $x = a$ if:

    i) $f(a)$ is defined (exists)

    ii) $\lim_{x \to a} f(x)$ exists, and

    iii) $\lim_{x \to a} f(x) = f(a)$

    Otherwise, f is discontinuous at $x = a$.

    $\lim_{x \to a} f(x)$ is finite if and only if both corresponding one-sided limits are finite and are equal — that is,

    $$\lim_{x \to a} f(x) = L \iff \lim_{x \to a^+} f(x) = L = \lim_{x \to a^-} f(x)$$

2. <u>Intermediate Value Theorem</u>

    A function $y = f(x)$ that is continuous on a closed interval $[a, b]$ takes on every value between $f(a)$ and $f(b)$.

    <u>Note</u>: If f is continuous on $[a, b]$ and $f(a)$ and $f(b)$ differ in sign, then the equation $f(x) = 0$ has at least one solution in the open interval (a, b).

3. <u>Limits of Rational Functions as $x \to \pm\infty$</u>

    1. $\lim_{x \to \pm\infty} \dfrac{f(x)}{g(x)} = 0$ if the degree of $f(x)$ < the degree of $g(x)$

        <u>Example:</u> $\lim_{x \to \infty} \dfrac{x^2 - 2x}{x^3 + 3} = 0$

    2. $\lim_{x \to \pm\infty} \dfrac{f(x)}{g(x)}$ is infinite if the degree of $f(x)$ > the degree of $g(x)$

        <u>Example:</u> $\lim_{x \to +\infty} \dfrac{x^3 + 2x}{x^2 - 8} = \infty$

    3. $\lim_{x \to \pm\infty} \dfrac{f(x)}{g(x)}$ is finite if the degree of $f(x)$ = the degree of $g(x)$

    <u>Note</u>: The limit will be the ratio of the leading coefficients of $f(x)$ and $g(x)$.

    <u>Example:</u> $\lim_{x \to \infty} \dfrac{2x^2 - 3x + 2}{10x - 5x^2} = -\dfrac{2}{5}$

4. <u>Horizontal and Vertical Asymptotes</u>

    1. A line $y = b$ is a <u>horizontal asymptote</u> of the graph of $y = f(x)$ if either $\lim_{x \to \infty} f(x) = b$ or $\lim_{x \to -\infty} f(x) = b$.

    2. A line $x = a$ is a <u>vertical asymptote</u> of the graph of $y = f(x)$ if either $\lim_{x \to a^+} f(x) = \pm\infty$ or $\lim_{x \to a^-} f(x) = \pm\infty$.

5.   <u>Average and Instantaneous Rate of Change</u>

    1.   <u>Average Rate of Change</u>: If (x_0, y_0) and (x_1, y_1) are points on the graph of $y = f(x)$, then the average rate of change of y with respect to x over the interval $[x_0, x_1]$ is

$$\frac{f(x_1) - f(x_0)}{x_1 - x_0} = \frac{y_1 - y_0}{x_1 - x_0} = \frac{\Delta y}{\Delta x}.$$

    2.   <u>Instantaneous Rate of Change</u>: If (x_0, y_0) is a point on the graph of $y = f(x)$, then the instantaneous rate of change of y with respect to x at x_0 is $f'(x_0)$.

6.   <u>Definition of the Derivative</u>

$$f'(x) = \lim_{h \to 0} \frac{f(x + h) - f(x)}{h} \text{ or } f'(a) = \lim_{x \to a} \frac{f(x) - f(a)}{x - a}$$

The latter definition of the derivative is the instantaneous rate of change of $f(x)$ with respect to x at $x = a$.

Geometrically, the derivative of a function at a point is the slope of the tangent line to the graph of the function at that point.

7.   <u>The Number e as a limit</u>

    1. $\lim\limits_{n \to +\infty} \left(1 + \dfrac{1}{n}\right)^n = e$

    2. $\lim\limits_{n \to 0} \left(1 + n\right)^{\frac{1}{n}} = e$

8.   <u>Rolle's Theorem</u>

If f is continuous on $[a, b]$ and differentiable on (a, b) and $f(a) = f(b)$, then there is at least one number c in the open interval (a, b) such that $f'(c) = 0$.

9.   <u>Mean Value Theorem</u>

If f is continuous on $[a, b]$ and differentiable on (a, b), then there is at least one number c in (a, b) such that $\dfrac{f(b) - f(a)}{b - a} = f'(c)$.

10.   <u>Extreme Value Theorem</u>

If f is continuous on a closed interval $[a, b]$, then $f(x)$ has both an absolute maximum and an absolute minimum on $[a, b]$.

11.   To find the maximum and minimum values of a function $y = f(x)$, locate

    1.   the point(s) where $f'(x)$ changes sign. To find the candidates first find where $f'(x) = 0$ or is infinite or does not exist.

    2.   the end points, if any, on the domain of $f(x)$.

Note: These are the only candidates for the value of x where $f(x)$ may have a maximum or a minimum.

12. Let f be differentiable for $a < x < b$ and continuous for $a \le x \le b$.

1. If $f'(x) > 0$ for every x in (a, b), then f is increasing on $[a, b]$.

2. If $f'(x) < 0$ for every x in (a, b), then f is decreasing on $[a, b]$.

13. Suppose that $f''(x)$ exists on the interval (a, b).

1. If $f''(x) > 0$ in (a, b), then f is concave upward in (a, b).

2. If $f''(x) < 0$ in (a, b), then f is concave downward in (a, b).

To locate the points of inflection of $y = f(x)$, find the points where $f''(x) = 0$ or where $f''(x)$ fails to exist. These are the only candidates where $f(x)$ may have a point of inflection. Then test these points to make sure that $f''(x) < 0$ on one side and $f''(x) > 0$ on the other.

14. If a function is differentiable at a point $x = a$, it is continuous at that point. The converse is false, i.e. continuity does <u>not</u> imply differentiability.

15. Local Linearity and Linear Approximation

The linear approximation of $f(x)$ near $x = x_0$ is given by $y = f(x_0) + f'(x_0)(x - x_0)$.

To estimate the slope of a graph at a point – draw a tangent line to the graph at that point. Another way is (by using a graphing calculator) to "zoom in" around the point in question until the graph "looks" straight. This method almost always works. If we "zoom in" and the graph looks straight at a point, say $x = a$, then the function is <u>locally linear</u> at that point.

The graph of $y = |x|$ has a sharp corner at $x = 0$. This corner cannot be smoothed out by "zooming in" repeatedly. Consequently, the derivative of $|x|$ does not exist at $x = 0$; hence, the function is <u>not</u> locally linear at $x = 0$.

16. Dominance and Comparison of Rates of Change

Logarithm functions grow slower than any power function (x^n).

Among power functions, those with higher powers grow faster than those with lower powers.

All power functions grow slower than any exponential function $(a^x, a > 1)$.

Among exponential functions, those with larger bases grow faster than those with smaller bases.

We say, that as $x \to \infty$:

1. $f(x)$ grows <u>faster</u> than $g(x)$ if $\lim\limits_{x \to \infty} \dfrac{f(x)}{g(x)} = \infty$ or if $\lim\limits_{x \to \infty} \dfrac{g(x)}{f(x)} = 0$.
    If $f(x)$ grows faster than $g(x)$ as $x \to \infty$, then $g(x)$ grows <u>slower</u> than $f(x)$ as $x \to \infty$.

2. $f(x)$ and $g(x)$ grow at the <u>same</u> rate as $x \to \infty$ if $\lim\limits_{x \to \infty} \dfrac{f(x)}{g(x)} = L \ne 0$ (L is finite and nonzero).

For example,

1. e^x grows faster than x^3 as $x \to \infty$ since $\lim\limits_{x \to \infty} \dfrac{e^x}{x^3} = \infty$

16.　Dominance and Comparison of Rates of Change (continued)

    2.　x^4 grows faster than $\ln x$ as $x \to \infty$ since $\lim\limits_{x \to \infty} \dfrac{x^4}{\ln x} = \infty$

    3.　$x^2 + 2x$ grows at the same rate as x^2 as $x \to \infty$ since $\lim\limits_{x \to \infty} \dfrac{x^2 + 2x}{x^2} = 1$

To find some of these limits as $x \to \infty$, you may use the graphing calculator. Make sure that an appropriate viewing window is used.

17.　Inverse Functions

    1.　If f and g are two functions such that $f(g(x)) = x$ for every x in the domain of g, and, $g(f(x)) = x$, for every x in the domain of f, then, f and g are inverse functions of each other.

    2.　A function f has an inverse function if and only if no horizontal line intersects its graph more than once.

    3.　If f is either increasing or decreasing in an interval, then f has an inverse function over that interval.

    4.　If f is differentiable at every point on an interval I, and $f'(x) \neq 0$ on I, then $g = f^{-1}(x)$ is differentiable at every point of the interior of the interval $g'(f(x)) = \dfrac{1}{f'(x)}$.

18.　Properties of $y = e^x$

    1.　The exponential function $y = e^x$ is the inverse function of $y = \ln x$.

    2.　The domain is the set of all real numbers, $-\infty < x < \infty$.

    3.　The range is the set of all positive numbers, $y > 0$.

    4.　$\dfrac{d}{dx}(e^x) = e^x$.

    5.　$e^{x_1} \cdot e^{x_2} = e^{x_1 + x_2}$

    6.　$y = e^x$ is continuous, increasing, and concave up for all x.

    7.　$\lim\limits_{x \to +\infty} e^x = +\infty$ and $\lim\limits_{x \to -\infty} e^x = 0$.

    8.　$e^{\ln x} = x$, for $x > 0$; $\ln(e^x) = x$ for all x.

19.　Properties of $y = \ln x$

    1.　The natural logarithm function $y = \ln x$ is the inverse of the exponential function $y = e^x$.

    2.　The domain of $y = \ln x$ is the set of all positive numbers, $x > 0$.

    3.　The range of $y = \ln x$ is the set of all real numbers, $-\infty < y < \infty$.

    4.　$y = \ln x$ is continuous, increasing, and concave down everywhere on its domain.

    5.　$\ln(ab) = \ln a + \ln b$

    6.　$\ln(a/b) = \ln a - \ln b$

    7.　$\ln a^r = r \ln a$

19.   Properties of $y = \ln x$ (continued)

   8. $y = \ln x < 0$ if $0 < x < 1$.

   9. $\lim\limits_{x \to +\infty} \ln x = +\infty$ and $\lim\limits_{x \to 0^+} \ln x = -\infty$.

   10. $\log_a x = \dfrac{\ln x}{\ln a}$

20.   L'Hôpital's Rule

   If $\lim\limits_{x \to a} \dfrac{f(x)}{g(x)}$ is of the form $\dfrac{0}{0}$ or $\dfrac{\infty}{\infty}$, and if $\lim\limits_{x \to a} \dfrac{f'(x)}{g'(x)}$ exists, then $\lim\limits_{x \to a} \dfrac{f(x)}{g(x)} = \lim\limits_{x \to a} \dfrac{f'(x)}{g'(x)}$.

21.   Trapezoidal Rule

   If a function f is continuous on the closed interval $[a, b]$ where $[a, b]$ has been partitioned into n subintervals $[x_0, x_1], [x_1, x_2], \ldots, [x_{n-1}, x_n]$, each of length $(b - a)/n$, then

   $$\int_a^b f(x)\,dx \approx \frac{b - a}{2n}[f(x_0) + 2f(x_1) + 2f(x_2) + \ldots + 2f(x_{n-1}) + f(x_n)].$$

   The Trapezoidal Rule is the average of the left-hand and right-hand Riemann sums.

22.   Properties of the Definite Integral

   Let $f(x)$ and $g(x)$ be continuuous on $[a, b]$.

   1. $\displaystyle\int_a^b c \cdot f(x)\,dx = c\int_a^b f(x)\,dx$, c is a nonzero constant.

   2. $\displaystyle\int_a^a f(x)\,dx = 0$

   3. $\displaystyle\int_b^a f(x)\,dx = -\int_a^b f(x)\,dx$

   4. $\displaystyle\int_a^b f(x)\,dx = \int_a^c f(x)\,dx + \int_c^b f(x)\,dx$, where f is continuous on an interval containing the numbers a, b, and c, regardless of the order a, b, and c.

   5. If $f(x)$ is an odd function, then $\displaystyle\int_{-a}^a f(x)\,dx = 0$

   6. If $f(x)$ is an even function, then $\displaystyle\int_{-a}^a f(x)\,dx = 2\int_0^a f(x)\,dx$

   7. If $f(x) \geq 0$ on $[a, b]$, then $\displaystyle\int_a^b f(x)\,dx \geq 0$

   8. If $g(x) \geq f(x)$ on $[a, b]$, then $\displaystyle\int_a^b g(x)\,dx \geq \int_a^b f(x)\,dx$

23. <u>Definition of the Definite Integral as the Limit of a Sum</u>

    Suppose that a function $f(x)$ is continuous on the closed interval $[a, b]$. Divide the interval into n equal subintervals, of length $\Delta x = \dfrac{b-a}{n}$. Choose one number in each subinterval i.e. x_1 in the first, x_2 in the second, $\ldots, x_k$ in the kth, $\ldots$, and x_n in the nth. Then $\displaystyle\lim_{n\to\infty} \sum_{k=1}^{n} f(x_k)\Delta x = \int_a^b f(x)\ dx$.

24. <u>Fundamental Theorem of Calculus</u>

    $$\int_a^b f(x)\ dx = F(b) - F(a)\,, \text{ where } F'(x) = f(x)$$

    $$\frac{d}{dx}\int_a^x f(t)\ dt = f(x)$$

    $$\frac{d}{dx}\int_a^{g(x)} f(t)\ dt = f(g(x))\ g'(x)\,.$$

25. <u>Velocity, Speed, and Acceleration</u>

    1. The <u>velocity</u> of an object tells how fast it is going <u>and</u> in which direction. Velocity is an instantaneous rate of change.

    2. The <u>speed</u> of an object is the absolute value of the velocity, $|v(t)|$. It tells how fast it is going disregarding its direction.

       The speed of a particle <u>increases</u> (speeds up) when the velocity and acceleration have the same signs. The speed <u>decreases</u> (slows down) when the velocity and acceleration have opposite signs.

    3. The <u>acceleration</u> is the instantaneous rate of change of velocity – it is the derivative of the velocity – that is, $a(t) = v'(t)$. Negative acceleration (deceleration) means that the velocity is decreasing. The acceleration gives the rate at which the velocity is changing.

    Therefore, if x is the displacement of a moving object and t is time, then:

    i)   velocity $= v(t) = x'(t) = \dfrac{dx}{dt}$

    ii)   acceleration $= a(t) = x''(t) = v'(t) = \dfrac{dv}{dt} = \dfrac{d^2x}{dt^2}$

    iii) $v(t) = \displaystyle\int a(t)\ dt$

    iv) $x(t) = \displaystyle\int v(t)\ dt$

    <u>Note</u>: The <u>average</u> velocity of a particle over the time interval from t_0 to another time t, is

    $$\text{Average Velocity} = \frac{\text{Change in position}}{\text{Length of time}} = \frac{s(t) - s(t_0)}{t - t_0}\,,$$ where $s(t)$ is the position of the particle at time t.

26. The average value of $f(x)$ on $[a, b]$ is $\dfrac{1}{b-a}\displaystyle\int_a^b f(x)\ dx$.

27.   <u>Area Between Curves</u>

If f and g are continuous functions such that $f(x) \geq g(x)$ on $[a, b]$, then the area between the curves is $\int_a^b [f(x) - g(x)] \, dx$.

28.   <u>Integration By Parts</u>

If $u = f(x)$ and $v = g(x)$ and if $f'(x)$ and $g'(x)$ are continuous, then $\int u \, dv = uv - \int v \, du$.

<u>Note</u> : The goal of the procedure to to choose u and dv so that $\int v \, du$ is easier to integrate than the original problem.

<u>Suggestion</u> :

When "choosing" u, remember the acronym L.I.A.T.E., where L is the logarithmic function, I is an inverse trigonometric function, A is an algebraic function, T is a trigonometric function, and E is the exponential function. Just choose u as the first expression in L.I.A.T.E. (and dv will be the remaining part of the integrand). For example, when integrating $\int x \ln x \, dx$, choose $u = \ln x$, since L comes first in L.I.A.T.E., and $dv = x \, dx$. When integrating $\int x e^x dx$, choose $u = x$, since x is an algebraic function, and A comes before E in L.I.A.T.E., and $dv = e^x \, dx$. One more example, when integrating $\int x \operatorname{Arctan} x \, dx$, let $u = \operatorname{Arctan} x$, since I comes before A in L.I.A.T.E., and $dv = x \, dx$.

29.   <u>Volume of Solids of Revolution</u>

Let f be nonnegative and continuous on $[a, b]$, and let R be the region bounded above by $y = f(x)$, below by the x-axis, and on the sides by the lines $x = a$ and $x = b$.

When this region R is revolved about the x-axis, it generates a solid (having circular cross sections) whose volume $V = \int_a^b \pi [f(x)]^2 \, dx$.

30.   <u>Volumes of Solids with Known Cross Sections</u>

1.   For cross sections of area $A(x)$, taken perpendicular to the x-axis, volume $= \int_a^b A(x) \, dx$.

2.   For cross sections of area $A(y)$, taken perpendicular to the y-axis, volume $= \int_c^d A(y) \, dy$.

31.   <u>Solving Differential Equations: Graphically and Numerically</u>

<u>Slope Fields</u>

At every point (x, y) a differential equation of the form $\frac{dy}{dx} = f(x, y)$ gives the slope of the member of the family of solutions that contains that point. A slope field is a graphical representation of this family of curves. At each point in the plane, a short segment is drawn whose slope is equal to the value of the derivative at that point. These segments are tangent to the solution's graph at the point.

The slope field allows you to sketch the graph of the solution curve even though you do not have its equation. This is done by starting at any point (usually the point given by the initial condition), and moving through the segments in the direction they indicate.

Some calculators have built in operations for drawing slope fields; for calculators without this feature there are programs available for drawing them.

31. <u>Solving Differential Equations: Graphically and Numerically - (continued)</u>

<u>Euler's Method</u>

Euler's Method is a way of approximating points on the solution of a differential equation $\frac{dy}{dx} = f(x,y)$. The calculation uses the tangent line approximation to calculate the approximate coordinates of the points of the solution function. Starting with the given point – the initial condition – (x_1, y_1) and a small step size, Δx, use the formulas $x_{n+1} = x_n + \Delta x$ and $y_{n+1} = y_n + f'(x_n, y_n)\Delta x$ to calculate succeeding points. The accuracy of the method decreases with larger values of Δx. The error increases as each successive point is used to find the next. Calculator programs are available for doing this calculation.

32. <u>Definition of Arc Length</u>

If the function given by $y = f(x)$ represents a smooth curve on the interval $[a, b]$, then the arc length of f between a and b is given by $s = \int_a^b \sqrt{1 + [f'(x)]^2}\, dx$.

33. <u>Improper Integral</u>

$\int_a^b f(x)\, dx$ is an improper integral if

1. f becomes infinite at one or more points in the interval of integration, or

2. one or both of the limits of integration is infinite, or

3. both (1) and (2) hold.

34. <u>Parametric Form of the Derivative</u>

If a smooth curve C is given by the parametric equations $x = f(t)$ and $y = g(t)$, then the slope of the curve C at (x, y) is $\dfrac{dy}{dx} = \dfrac{dy}{dt} \div \dfrac{dx}{dt}, \dfrac{dx}{dt} \neq 0$.

<u>Note:</u> The second derivative, $\dfrac{d^2y}{dx^2} = \dfrac{d}{dx}\left[\dfrac{dy}{dx}\right] = \dfrac{d}{dt}\left[\dfrac{dy}{dx}\right] \div \dfrac{dx}{dt}$.

35. <u>Arc Length in Parametric Form</u>

If a smooth curve C is given by $x = f(t)$ and $y = g(t)$ and these functions have continuous first derivatives with respect to t for $a \leq t \leq b$, and if the point $P(x, y)$ traces the curve exactly once as t moves from $t = a$ to $t = b$, then the length of the curve is given by

$$s = \int_a^b \sqrt{\left(\frac{dx}{dt}\right)^2 + \left(\frac{dy}{dt}\right)^2}\, dt = \int_a^b \sqrt{[f'(t)]^2 + [g'(t)]^2}\, dt.$$

36.   Polar Coordinates

1.   <u>Cartesian *vs.* Polar Coordinates:</u> The polar coordinates (r, θ) are related to the Cartesian coordinates (x, y) as follows:

$$x = r \cos \theta \text{ and } r = \sin \theta;$$

$$\tan \theta = \frac{y}{x} \text{ and } x^2 + y^2 = r^2$$

2.   <u>Area in Polar Coordinates:</u> If f is continuous and nonnegative on the interval $[\alpha, \beta]$, then the area of the region bounded by the graph of $r = f(\theta)$ between the radial lines $\theta = \alpha$ and $\theta = \beta$ is given by

$$A = \frac{1}{2} \int_\alpha^\beta [f(\theta)]^2 \, d\theta = \frac{1}{2} \int_\alpha^\beta r^2 \, d\theta$$

37.   Sequences and Series

1.   If a sequence $\{a_n\}$ has a limit L, that is, $\lim\limits_{n \to \infty} a_n = L$, then the sequence is said to <u>converge</u> to L. If there is no limit, the sequence <u>diverges</u>. If the sequence $\{a_n\}$ converges, then its limit is unique. Keep in mind that $\lim\limits_{n \to \infty} \dfrac{\ln n}{n} = 0$; $\lim\limits_{n \to \infty} x^{\frac{1}{n}} = 1$; $\lim\limits_{n \to \infty} \sqrt[n]{n} = 1$; $\lim\limits_{n \to \infty} \dfrac{x^n}{n!} = 0$. These limits are useful and arise frequently.

2.   The harmonic series $\sum\limits_{n=1}^{\infty} \dfrac{1}{n}$ diverges; the geometric series $\sum\limits_{n=0}^{\infty} ar^n$ converges to $\dfrac{a}{1-r}$ if $|r| < 1$ and diverges if $|r| \geq 1$ and $a \neq 0$.

3.   The p-series $\sum\limits_{n=1}^{\infty} \dfrac{1}{n^p}$ converges if $p > 1$ and diverges if $p \leq 1$.

4.   <u>Limit Comparison Test:</u> Let $\sum\limits_{n=1}^{\infty} a_n$ and $\sum\limits_{n=1}^{\infty} b_n$ be a series of nonnegative terms, with $a_n \neq 0$ for all sufficiently large n, and suppose that $\lim\limits_{n \to \infty} \dfrac{b_n}{a_n} = c > 0$. Then the two series either both converge or both diverge.

5.   <u>Alternating Series:</u> Let $\sum\limits_{n=1}^{\infty} a_n$ be a series such that

i)   the series is alternating

ii) $|a_{n+1}| \leq |a_n|$ for all n or at least after a finite number of terms, and

iii) $\lim\limits_{n \to \infty} a_n = 0$

Then the series converges.

6.   The nth Term Test for Divergence: If $\lim\limits_{n \to \infty} a_n \neq 0$, then the series diverges.

Note that the converse is *false*, that is, if $\lim\limits_{n \to \infty} a_n = 0$, the series may or may not converge.

37.   Sequences and Series (continued)

7.   A series $\sum a_n$ is <u>absolutely convergent</u> if the series $\sum |a_n|$ converges. If $\sum a_n$ converges, but $\sum |a_n|$ does not converge, then the series is <u>conditionally convergent</u>. Keep in mind that if $\sum_{n=1}^{\infty} |a_n|$ converges, then $\sum_{n=1}^{\infty} a_n$ converges.

8.   <u>Comparison Test</u>: If $0 \leq a_n \leq b_n$ for all sufficiently large n, and $\sum_{n=1}^{\infty} b_n$ converges, then $\sum_{n=1}^{\infty} a_n$ converges. If $\sum_{n=1}^{\infty} a_n$ diverges, then $\sum_{n=1}^{\infty} b_n$ diverges.

9.   <u>Integral Test</u>: Let $f(x)$ be a positive, continuous, and decreasing function on $[1, \infty)$, and $a_n = f(n)$. The series $\sum_{i=1}^{\infty} a_n$ and the improper integral $\int_{1}^{\infty} a(n) \, dx$ both converge or both diverge.

10.   <u>Ratio Test</u>: Let $\sum a_n$ be a series with nonzero terms.

   i)   If $\lim\limits_{n \to \infty} \left| \dfrac{a_{n+1}}{a_n} \right| < 1$, then the series converges absolutely.

   ii)   If $\lim\limits_{n \to \infty} \left| \dfrac{a_{n+1}}{a_n} \right| > 1$, then the series is divergent.

   iii)   If $\lim\limits_{n \to \infty} \left| \dfrac{a_{n+1}}{a_n} \right| = 1$, then the test is inconclusive (and another test must be used).

11.   <u>Power Series</u>: A power series is a series of the form

$$\sum_{n=0}^{\infty} c_n x^n = c_0 + c_1 x + c_2 x^2 + \cdots + c_n x^n + \cdots \text{ or}$$

$$\sum_{n=0}^{\infty} c_n (x - a)^n = c_0 + c_1 (x - a) + c_2 (x - a)^2 + \cdots + c_n (x - a)^n + \cdots$$

in which the center a and the coefficients $c_0, c_1, c_2, \ldots, c_n, \ldots$ are constants.

The set of all numbers x for which the power series converges is called the <u>interval of convergence</u>.

37. <u>Sequences and Series (continued)</u>

12. <u>Taylor Series</u>: Let f be a function with derivatives of all orders throughout some interval containing a as an interior point. Then the Taylor series generated by f at a is

$$\sum_{k=0}^{\infty} \frac{f^{(k)}(a)}{k!}(x-a)^k = f(a) + f'(a)(x-a) + \frac{f''(a)}{2!}(x-a)^2 + \cdots + \frac{f^{(n)}(a)}{n!}(x-a)^n + \cdots$$

The remaining terms after the term containing the nth derivative can be expressed as a remainder to Taylor's Theorem:

$$f(x) = \sum_{i=1}^{n} \frac{f^{(n)}}{n!}(x-a)^n + R_n(x) \text{ where } R_n(x) = \frac{f^{(n+1)}(c)}{(n+1)!}(x-a)^{n+1} \text{ for some number}$$

c between x and a. The number $R_n(x)$ is called the Lagrange Form of the Remainder. $|R_n(x)|$ can be used to estimate the size of the error when the finite Taylor Polynomial is used to approximate the series. The series will converge for all values of x for which the remainder approaches zero as $n \to \infty$.

Lagrange's form of the remainder: $R_n x = \dfrac{f^{(n+1)}(c)(x-a)^{n+1}}{(n+1)!}$, where c lies between x and a.

The series will converge for all values of x for which the remainder approaches zero as $x \to \infty$.

13. <u>Frequently Used Series and their Interval of Convergence</u>

$$\frac{1}{1-x} = 1 + x + x^2 + \cdots + x^n + \cdots = \sum_{n=0}^{\infty} x^n, \quad |x| < 1$$

$$\sin x = x - \frac{x^3}{3!} + \frac{x^5}{5!} - \cdots + (-1)^n \frac{x^{2n+1}}{(2n+1)!} + \cdots = \sum_{n=0}^{\infty} \frac{(-1)^n x^{2n+1}}{(2n+1)!}, \quad |x| < \infty$$

$$\cos x = 1 - \frac{x^2}{2!} + \frac{x^4}{4!} - \cdots + \frac{(-1)^n x^{2n}}{(2n)!} + \cdots = \sum_{n=0}^{\infty} \frac{(-1)^n x^{2n}}{(2n)!}, \quad |x| < \infty$$

$$e^x = 1 + x + \frac{x^2}{2!} + \frac{x^3}{3!} + \cdots + \frac{x^n}{n!} + \cdots = \sum_{i=1}^{n} \frac{x^n}{n!}, |x| < \infty$$

INDEX
(For Multiple-Choice Questions)

Roman numerals in boldface type represent the sample exam numbers; other numbers are question numbers.